APPLES & ORCHARDS IN SUSSEX

Apples & Orchards in Sussex

Brian Short

with

Peter May ⁂ Gail Vines ⁂ Anne-Marie Bur

Action in rural Sussex

Brighton Permaculture Trust

ACTION IN RURAL SUSSEX & BRIGHTON PERMACULTURE TRUST

APPLES & ORCHARDS IN SUSSEX

First published in the United Kingdom by:
Action In Rural Sussex and Brighton Permaculture Trust
Sussex House
212 High Street
Lewes
East Sussex
BN7 2NH

Design: Peter Flanagan
Original Illustration: Michael Munday

Printed in the UK using FSC paper & vegetable-based inks containing no heavy metal, by Calico, West Sussex, 01273 833353

ISBN 978-1-873850-23-7

Contents

Foreword

What a great book - the editors and all the contributors should be congratulated. I am fascinated by the many apples stories, which cover the topic from field to fork. In an age when the global food market and supply system dominates, this books shows how we can re-engage with local food and even re-train our taste buds. Apples have a long culinary history; there are written recipes for apple dishes dating back to the thirteenth and fourteenth centuries. Apples have been grown in northern Europe from time immemorial, and varieties have been developed to suit local climates, such as the Russian apple White Transparent which ripens quickly in northern climes with shorter summers. Such developments have long been the preserve of orchardists, and Sussex has a long tradition in this respect. Even a globally popular apple today, Granny Smith, famously has Sussex connections, although it cannot be grown in this country. Marie Ann Smith from Peasmarsh in Sussex moved to Australia and established orchards in the area round around Sydney, where she became known as Granny. Then in 1868 she discovered a small seedling bearing a few apples. The story is that this was the result of her bringing back from market a case of rotting Tasmanian apples – probably French Crabs – and hence Granny Smiths were "born". All with a trail back to Sussex.

Many of the more recent activities described in this book were jump-started in 2003 at a conference in Lewes looking at local food initiatives. I was there at the birth or "grafting", and nine years on it is amazing what has been achieved. I am interested in food policy and the current book shows how food policy can be developed, through the linking of history, family relations, health, ecology, growing, tasting, cooking and above all enjoyment. The lesson is simple: engage people with food and the growing of it, rather than merely engaging with the problem at a conceptual level. The current work shows the way forward for many other projects.

So happy and informative reading, and I look forward to writing the book foreword for the 20th anniversary of this work.

Martin Caraher
Professor of Food and Health Policy
Centre for Food Policy
City University, London
Thinker in Residence
Deakin University, Melbourne (2012)

How this book began…

In 2004, Action In Rural Sussex, a charity that supports rural communities, began planting school orchards with horticulturist Peter May, and quickly joined forces with another orchard expert too, Bryn Thomas of Brighton Permaculture Trust. We taught children how to plant fruit trees and ran fruit cookery sessions with them often led by Community Chef Robin van Creveld. The creation of school orchards was one response amongst many at the time to concerns about children's diets, and our growing cultural and economic distance from local food sources.

We started to investigate local Sussex apple varieties to plant in school grounds and quickly discovered that some were very rare and most were only known to exist with any certainty in the National Fruit Collection or West Dean Gardens. As we began to propagate and plant Sussex apple trees the promotion of biodiversity emerged as another theme of our work and we embarked on a plan to plant them all in the community-run orchards at Stanmer Park near Brighton, managed by Brighton Permaculture Trust. Requests for assistance with planning community orchards in other settings began to grow.

Along the way we had conversations with people across Sussex who regretted the loss of orchards that had been familiar features in their landscapes and lives. In decades past these orchards had supplied fruit to local greengrocers and provided work for many particularly at harvest time. We were made aware too of just how common it once was to have an apple tree or two in gardens and allotments and were struck by how many people used to have access to locally grown apples and the fond memories they evoked. The idea of a more concerted effort to uncover our orchard past took hold.

Historical research would help to fill in gaps in our knowledge: When did apple growing start? When did we grow most apples? How did all these Sussex varieties come about? Why have so many orchards disappeared? - and help us to get a fuller, richer idea of how orchards featured at different times in our history. But we also wanted to get a sense of how our current work might take heed of what has gone before – and not just from the large commercial orchards but from the small orchards that were scattered across our landscape over many centuries. Are there elements of how orchards were grown and fruit was used in the past that we could emulate? And as we move towards an uncertain future how can we adapt and move on from the past? Can community-scale interventions have an impact on our food supply, in urban as well as rural settings, and can we do this in a way that is friendly to wildlife too?

That's how this book began…

Anne-Marie Bur, Action In Rural Sussex

I. Ripe for Revival

1

Why apples in Sussex are ripe for revival

Today, most apples eaten in Sussex are bought in supermarkets, having travelled many, many miles to reach their consumers. Yet this overwhelming reliance on far-away growers is a surprisingly recent development. Apple trees thrive in many places in Sussex, and not so long ago locally-grown apples, as well as home-produced cider, were commonplace. Thousands upon thousands of orchards – consisting of just a few trees or dozens, in town plots, country gardens and farmland, listed in old wills, cited in legal disputes, delineated on old estate plans and parish maps – all testify to Sussex's long if neglected history of orcharding. These are special places, unique blends of wildness and cultivation.

Few trees are so attractive and at the same time so beneficial to a remarkable diversity of living creatures, not least us. Like their close relative our native wild crab, apple trees are festooned in spring with beautiful pink and white blossom, with each variety having its own distinctive hues and its own characteristic time of flowering. Each flower is simple, but perfectly designed: members of the rose family, apples remain faithful to an ancient floral shape that evolved to welcome the bees, hoverflies, butterflies and moths that pollinate the flowers, so enabling these alluring blooms to transform themselves into fruit.

Beautiful and useful: since ancient times, apple trees have been cultivated to delight the eye as much as the palate

Apples trees show an exceptional generosity: simply take a little care in planting and looking after the young tree, and it will reward you, year after year, with a rich harvest of delicious food, there for the picking. It is the easiest food we can grow in these islands, and is, nutritionally, of the highest quality. And almost any urban plot can find space for at least one apple tree, as Sussex folk once knew – just a few decades ago, Sussex, town and country, was awash in home-grown apples.

Delighting in this rich inheritance, Sussex is home to a host of unique varieties – apples full of character and intensely seasonal, with intriguing shapes, aromas, flavours and textures. Each variety – almost 30 still survive in cultivation – has its story of discovery and celebration. *Alfriston*, for instance – with its roots in Uckfield soil, despite its name – is such a good keeper that it was once acclaimed as the "sailors' apple". Stowed away on ships, it could stay wholesome for months on end, providing vitamin C to help fend off scurvy all the while. Meanwhile *Forge*, still found growing round the old iron workings of East Grinstead, was dubbed the "cottager's apple" not least because it is good for everything – eating, cooking and making cider.

Every apple variety has its own special qualities. Hawkridge, first described in Sussex in Victorian times, has sweet, juicy and soft fruit

One Sussex variety, *Egremont Russet* from the Petworth estate, deservedly has many admirers, but Sussex has also given the world many other worthy apples, aptly named, including *Crawley Beauty, Dr Hogg, Duck's Bill, Eastbourne Pippin, Edmund Jupp, First and Last, Hawkridge, Knobby Russet, Mannington's Pearmain, Saltcote Pippin, Sussex Mother, Tinsley Quince* and *Wadhurst Pippin* (for more about Sussex varieties, see chapter 2).

Why so many sorts?

Each variety of apple has its special qualities and has long been valued because of them. *Aldwick Beauty*, for instance, ripens in August, and is best eaten soon, while *Saltcote Pippin*, picked in October, is good off the tree but becomes sweeter and more aromatic after a month or so in storage. *Dr Hogg* makes fabulous baked apples, while *Crawley Beauty* cooks to a beautiful puree. Some apples were ideal for dumplings, others were just the thing for mincemeat. Perhaps no fruit is more various that the English apple.

Not so long ago, knowledge of varieties and their special qualities was "part of the inheritance of the country-bred man" (and woman!), but urbanisation and global markets have eroded that once taken-for-granted birthright. Even before the Second World War, commentators were regretting its loss, "because this lack of knowledge favours the flood of 'fashionless' foreign apples which swamp our markets, mass-produced and looking almost machine-made in their bright polish and shop finish."[1] Today, Gala is the best-selling British-grown variety, while sales of Cox continue to decline. Gala's bland sweetness will seem dull to anyone who has had the chance to taste any of the renowned old English varieties, with their delicious balance of acidity and sugars. But Gala is a very pretty red apple and these days, fruit wholesalers supplying the nation's fruit bowls in schools, hotels, offices and canteens seem largely to trade in "red apples or green apples," and on price rather than taste, says food writer Hattie Ellis in *What To Eat?*[2] "It was almost as if the apples were not there to be eaten." And of course many of them won't be – on average, people throw away a third of the food they buy, Hattie points out. Could it be that by simply buying good, wholesome food, fruit and veg, we feel we have done our duty and don't bother to eat it? But which came first, she wonders, the poor apples or the wastefulness? "Perhaps we buy and throw away for a good reason: the apples don't taste good. Put like that, is a good apple really so expensive?" A taste odyssey awaits anyone able to seek out the old English varieties that still turn up at farmers' markets and farm shops and local enterprises across Sussex.

Apples do particularly well in Britain, a happy circumstance that has inspired many a patriotic outburst over the years. "To an Englishman the apple is the fruit of fruits," trumpeted H.V. Taylor in 1936

Local varieties are worth eating and growing too because they thrive on their native soils, and may even be less prone to disease when grown in their original place. *Wadhurst Pippin*, for instance, thrives on the challenging clay soils of its homeland in the High Weald. Many Sussex apples grew (and often still do) tucked away in small home orchards, school and hospital grounds, estate gardens and commercial plantations. Scattered along field margins and road verges too are "wildings" – genetically novel apple trees that no one has intentionally planted. Sprouted from discarded apple cores, these trees grow happily in neglected and abandoned places,

Fruit appreciation: Community Chef Robin van Creveld contrasts global with local at an apple cookery lesson with pupils of Manor Primary School, Uckfield

and sometimes produce edible, even tasty, fruit. Many celebrated varieties – including Sussex's own *Mannington's Pearmain* – have arisen in this way, springing into growth from cider makers' apple pressings dumped under a hedge or in the corner of a field. Such so-called "seedling" apples growing on their own roots can then be propagated *ad infinitum* from cuttings grafted onto rootstocks. But new varieties are always waiting to emerge from the apple pips' storehouse of genetic diversity. Indeed, a new Sussex apple, grown from a supermarket apple and discovered some 30 years ago by Anne Markwick and transplanted to the Plumpton College site at Stanmer Park, looks set to gain official recognition as the Stanmer Pippin. And there's another seedling tree with tasty red-skinned fruit, growing on a chalky Brighton allotment, that just might one day be hailed as the Patcham Pippin. Across the county, there must be hundreds of seedling apple trees alive today, and who knows which might yet win recognition as a worthy Sussex variety born in the twenty-first century.[3]

Apples have been avidly cultivated in Sussex for hundreds of years, not least for their obliging readiness to ferment into an alcoholic beverage – traditionally either cider, with a strength similar to beer and ale, or cyder, a much stronger drink on a par with wine. Far more than just a refreshing tipple, it was a kind of currency too – wages, debts, tithes and rents, all were widely paid in cider. Cider making in Sussex has followed the "eastern tradition" – made, as in Suffolk and Norfolk, from dessert and culinary apples rather than varieties grown exclusively for cider.

Once it paid the rent: this Sussex cider press is now a museum piece in Amberley, but not so long ago, when cider functioned as a kind of currency, such equipment was highly prized

Cider has had an image problem – not least because much pub and supermarket fare is industrially made from imported apple juice – and the craft of making good cider largely died out years ago under pressure from imported wine and local beer. But in the broadly self-sufficient land economies of the past, cider offered many advantages. Beer and ale require large amounts of good barley that might otherwise feed cattle or labourers, as well as fuel to boil the mash and malt the grain. By contrast, cider, traditionally made, has impeccable ecological credentials. To make this eminently pleasing drink, "apples only need to be gathered and pressed – nature does the rest", as Somerset cider-maker and Apple Day poet James Crowden points out.[4] What's more, the crushed apple waste, or pomace, can be fed to livestock in late autumn, just as grassland forage is in steep decline. Cider is the perfect drink for the Sussex Weald, where barley won't grow but apples thrive, and there are encouraging signs of a revival in the fortunes and quality of locally-made cider.

Cider offered health benefits too. Rich in vitamin C, the fresh, unpasteurised cider or cyder drunk in the eighteenth century helped to stave off sailors' scurvy, while the fermenting yeast made these untreated brews rich in vitamin B_{12} as well. Today, nutritionists hold apples in high regard, being rich in vitamins, potassium salts, antioxidant flavonoids, carotenoids and fruit fibre – which helps to ensure a slow

Fruitful future: Bryn Thomas of Brighton Permaculture Trust has helped to plant hundreds of young apple trees in school and community orchards across Sussex – here on a follow up visit at St. Luke's church, Brighton with their Orchard Leader Abi

release of sugars into the bloodstream and create a feeling of satiety from just 40 calories. Research links an apple-rich diet with lower risks of heart disease, cancer, diabetes and asthma. Locally-grown fruit might even help us keep trim. Psychologist Andrew Hill at the Institute of Health Sciences at the University of Leeds suspects that the rising tide of obesity can be fought in part through forging more connections with our food, by growing our own and searching out local producers.[5]

Co-evolution at work

Orchards are places where people and nature can come together in creative and productive ways – in fact, they're one of the best exemplars we have of people and nature working in harmony. "Orchards are a halfway house, neither quite field nor woodland but something in the middle," James Crowden writes. Apples and people have evolved together over thousands of years, not least in Sussex, where apple orchards have a long and productive history that is far from over. Many different parts of Sussex – from the eastern Weald to the west near Petworth, the coastal plain and the foothills and valleys of the downs – have fostered substantial swathes of orchard land. For centuries, orchards thrived in the corners of gardens and small holdings. Across the county, small-scale, diversified farms supported a variety of enterprises that were mutually supporting and which spread the risk of failure at harvest time – an eminently sustainable approach revitalised today in the permaculture movement. The people who planted orchards took the long view – and they invested in the stable productivity of the land.

Now is the time to reclaim our legacy and "rewild" the gardens and paddocks and fields of Sussex – celebrating and protecting the old orchards that remain and planting new apple trees that offer us food virtually for free, while providing nooks and crannies for all manner of wildlife.

Already, the revival is under way. Consumed fresh, or made into cider or apple juice, Sussex apples are being enjoyed with an enthusiasm buoyed up by the rediscovery of ancient traditions of growing and working with local produce. Apple Day – launched on 21 October 1990 in London's Covent Garden by the environmental charity Common Ground, which sparked the current orchard revival – is now an autumnal

Fresh apples and apple preserves galore are on sale at the annual Apple Day at Stanmer Park near Brighton

fixture in many places in Sussex, including West Dean near Chichester and Stanmer Park near Brighton.[6] These annual events celebrate the wonderful diversity of apples and other tree fruit in Britain – variety that enriches our sense of place.

It's telling that one of the most popular features of many Apple Days is the apple identification service – upwards of 3000 varieties have been cultivated in Britain. Many apple varieties will grow well in Sussex, and increasingly people are finding space for an apple tree or two in their gardens or allotments – perhaps a Sussex variety, now available from specialist local growers (see chapter 12). Over the past few years, Action in rural Sussex and Brighton Permaculture Trust working together, have propagated over a thousand apple trees of mostly rare Sussex varieties, and more than sixty small orchards have been planted across Sussex – everywhere from

inner city estates to urban parks, village greens and primary schools, where pupils look after the trees and harvest the produce. Community groups throughout East and West Sussex are pioneering community orchards, planting fruit trees that are tended collectively and the harvest shared. These communal orchards are forging new social links and providing fresh fruit, juice and sometimes cider (see chapter 10).

Thanks to years of lobbying by Common Ground, nature conservationists too have recently recognised that orchards can be wonderful places for wild plants and animals, and deserve to be ranked among habitats awarded the highest conservation priority (see chapter 11). Recent aerial surveys undertaken by the People's Trust for Endangered Species reveal that Sussex as a whole still has more than 900 old orchards dotted about the countryside. In West Sussex in particular, local volunteers have begun to track down these remarkable survivors, as a spur to conservation initiatives aiming to link up isolated pockets of wildlife-friendly habitat. The West Weald Landscape project has found many small clusters of old apple trees hidden away in private rural gardens and old estate land.

In some of these as yet uncharted places, Sussex varieties now thought to be lost – among them *Petworth Non Pareil* – may yet be rediscovered. But no one should think of orchards as museum pieces, mere relics of the past. In the twenty-first century, nothing could be more relevant to current concerns. As a rich source of local food, orchards reduce food miles and promote food security. The local economy benefits too, as commercial producers, and even small-scale orchardists with a surplus, provide apples and juice to local greengrocers and farmers' markets.

Below: nature's tithe - jays enjoy eating apples too, and there's usually plenty to go round

Orchards do their bit to combat global warming too – as a glorious amalgam of trees and permanent grassland, they act as carbon sinks, tying up carbon dioxide in the trees, grasses, fungi, invertebrates and the soil itself. Trees as a source of food should be taken much more seriously, argues biologist Colin Tudge, as modern advocates of the ancient principles of "agroforestry" and "forest gardening" have long advocated (see chapter 11). Grains such as wheat, rye, maize and rice, dominate our diets through a mere quirk of history, he argues in *The Secret Life of Trees*.[7] Apple trees help to prevent soil erosion, help to keep the climate equitable, provide timber as well as food, and can readily be combined with grazing sheep, or with conventional crops grown in the gaps between the trees. Orchards can provide ideal forage too for free-range hens – which happily potter under the safe cover of trees – and are excellent places for honeybee hives. Making the most of what such productive trees can offer us "is one of the great hopes for the future," Tudge concludes. Apples can be grown alongside pears, plums – the Victoria plum has Sussex origins – and even figs – as the remnant of a once-substantial eighteenth-century fig garden at Tarring, near Worthing, testifies.

All the while they're growing food for us and for nature, the orchards themselves are good places for people. Community orchards and fruit-sharing schemes rebuild social connections and communities' sense of place, and simply give people somewhere pleasant to be out-of-doors. They might even help to fight "nature-deficit disorder" – a diffuse sense of unease arguably experienced by many urban dwellers, surrounded as we are by artificial environments and often utterly divorced from natural green spaces. In a famous study in the US from the 1980s, just the sight of a tree from their hospital bed made patients recover significantly more rapidly than those whose only view was a treeless urban scene.

Opposite: Pupils at Clymping Primary School harvest apples from trees they helped to plant and look after

Trees for life: community orchards can give urban families a chance to enjoy just mooching about in a cool green place, as here at Stanmer on a busy Apple Day

And no wonder. "We sometimes forget that ours is a cultural landscape," writes Sue Clifford of Common Ground. "It is our great creation: underpinned by nature, it is a physical thing and an invisible web."[8]

Sussex is apple country too!

This is a book about growing apples in Sussex. Other counties are more renowned for their orchards: not least neighbouring Kent, with its long-established market gardens within easy reach of London, and Herefordshire, Somerset, Devon and Gloucestershire in the west of England, where geology and climate conspire against arable farming. Yet Sussex, with its equitable climate and geological diversity, offers a host of apple-friendly microclimates. Stretching over 400 square kilometres, the county enjoys an extraordinary variety of landscapes, with nowhere more than 40 kilometres from the sea. Apples won't grow happily everywhere – the windy top of the chalky downs are a prime example – but the intimate and diverse landforms of Sussex give its orchards a special character and charm. With a wide but scattered distribution, orchards have thrived in quite different parts of the county, not least the coastal plains round Chichester, the greensand hills west of Pulborough, and most of all in the High Weald, a hilly country of mixed sands and clays (see chapter 2).

Over the centuries, at least ever since the Romans arrived, Sussex has proved to be prime territory for growing apples. Anglo-Saxon settlers probably established orchards, boosted by the arrival of the next wave of invaders from Normandy. And as feudal economies were gradually transformed into mercantile enterprises, the orchards of Sussex became key components in agricultural "improvements". But all the while, small orchards, or even just a tree or two, continued to be vital assets attached to private houses, schools, vicarages and hospitals, supplying wholesome and extremely varied fresh fruit – and the wherewithal to make cider.

Greengrocers in every Sussex town once provided ready markets for local fruit: Lewes, 1890s

As cheap, attractive and intensively-farmed apples grown in the US, Canada, South Africa and New Zealand began to swamp the British market, starting round about the 1890s, many wonderful locally-grown apples disappeared from our nation's greengrocers. Yet despite the pressure from imports, commercial orcharding in Sussex doubled in acreage from the 1870s to a high point in the mid 1950s. Since that peak, orchards have undergone a

catastrophic decline, as misguided EC and government policies put pressure on English farmers to grub up old orchards. But in recent years, there are signs of a modest upturn in the number of professional growers, combined with an encouraging revival of the small-scale, local orchards that were always the hallmark of the county's apples.

This celebration of Sussex apples and orchards hopes to do many things, but most of all to convey a sense of their significance over many centuries, as a spur to a resurgence in apple-growing in both the eastern and western parts of the historic county.[9] Apples can be grown on any scale – from a single old tree that has somehow survived in an urban carpark or a back garden to a commercial bush-plantation. Apple orchards can promote rural regeneration and local markets, and help communities to re-build social networks and a sense of knowing and valuing the place where they live. Their potential, as vehicles for social change and cohesion, for wildlife regeneration and rehabilitation, and for marrying the two as nature and culture are reunited, is as yet insufficiently realised. There's life beyond those apples trapped in the supermarkets' plastic bags.

Small can be beautiful: orchards come in many shapes and sizes, and with a bit of help can grow almost anywhere. One small urban patch of green in central Brighton - Phoenix Community Centre - now has its own mini-orchard

A job well done: pupils from St Luke's Primary School show off a tree they helped to plant on the Craven Vale estate in Brighton with members of the Community Association and the Local Fruit Futures team

Endnotes for Chapter 1

1 Foreword by Sir Daniel Hall, 1936, v, in HV Taylor's *Apples of England*, revised third edition, London: Crosby Lockwood, 1948.

2 Hattie Ellis, *What To Eat?* London: Portobello Books, 2012, 15.

3 Hundreds if not thousands of seedling trees grow in a few locations on the downs around Brighton, probably originating from pig farming in the 1950s when the pigs were fed on fruit waste. These trees provide an important food source for local people and may well yield up a number of new varieties suited to local conditions.

4 James Crowden, *Cider – the forgotten miracle*, Somerton, Somerset: Cyder Press 2, 1999, xiv.

5 Hattie Ellis, op. cit., 271.

6 Sue Clifford and Angela King, *The Common Ground Book of Orchards*, London: Common Ground, 2000.

7 Colin Tudge, *The Secret Life of Trees*, Harmondsworth: Penguin, 2005

8 www.commonground.org.uk

9 This book embraces the historic county of Sussex, an ancient unit of land administration dating back over a thousand years. The county was divided into East and West Sussex for most administrative business only comparatively recently, in 1865, with separate county councils established in 1888. The new unitary authority of Brighton and Hove was created in 1997.

2. Sussex Varieties

2

Local distinctiveness: a guide to Sussex apple varieties

"Why do we need so many kinds of apples? Because there are so many folks. A person has a right to gratify his legitimate taste. If he wants twenty or forty kinds of apples for his personal use…he should be accorded the privilege. There is merit in variety itself. It provides more contact with life, and leads away from uniformity and monotony."[1]

Good things can happen when we least expect them. A delicious Sussex apple variety, eventually named *Mannington's Pearmain*, for instance, grew up quite by chance round about 1770 from the leftover apple pressings – cider pomace – thrown under the hedge of a local blacksmith, Mr Turley of Uckfield. It got its name some eighty years later, when Mr Turley's grandson, one John Mannington, sent his apples to London to win official recognition for his grandfather's wonderful tree.

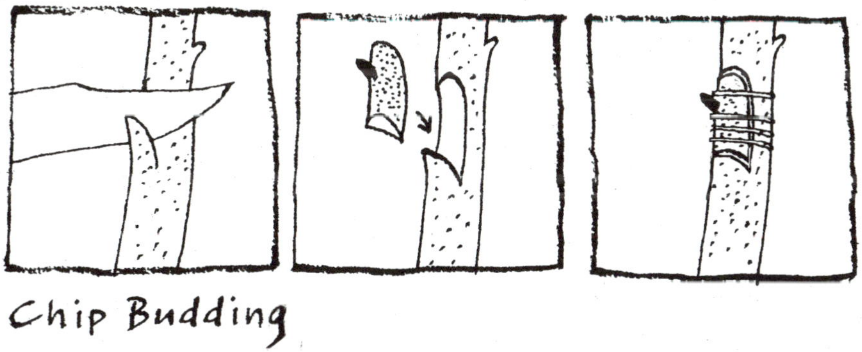

New apple varieties invariably spring from apple pips, which grow up as seedlings to become mature trees on their own roots, each with a new and unique complement of genes. The apple, like many other fruits, does not breed true from seed. A few pips, however, will strike lucky, and grow into hardy, healthy trees festooned with apples that people like. Any one tree has a limited lifespan – apples can live about a hundred years – but that tree can be propagated by taking cuttings (called scions) and grafting them onto rootstocks. The grafted tree will then grow up to be just like the original.

Facing page: four Sussex apples by Lewes-based botanical artist Nicky Ashworth. From the top clockwise: Saltcote Pippin, Golden Pippin, Tinsley Quince, Hawkridge

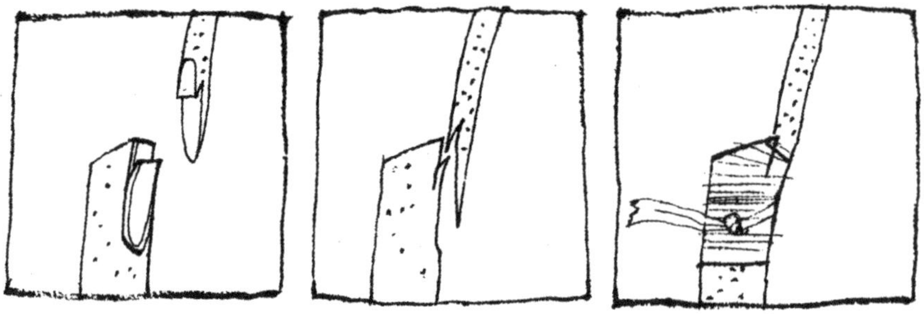

Whip & Tongue grafting

Two different grafting techniques are widely used today. One approach, called chip budding, uses a sharp knife to fit a chip of young applewood containing a bud onto a rootstock in August. Once tied tightly together, the chipbud and rootstock are left until the following spring. The rootstock is then cut back and the bud grows out to form a young tree by the autumn. In whip and tongue grafting, the scion or cutting is collected when the tree is dormant and grafted onto a rootstock in March. The scion and rootstock are tied tightly together with polythene tape. The graft will grow to form a young tree by autumn.

This ancient technique of grafting has enabled growers to maintain particular varieties of apples and other fruit over hundreds of years, with each grafted tree genetically identical to its parent. In traditional orchards, over many decades, trees most suited to local soils and micro-climatic conditions that yielded well would have been chosen to replace ageing trees, boosting the diversity of local apples and enhancing local distinctiveness.

Often, the origins of now-famous varieties remain obscure – for instance nobody knows who or what planted the seed that gave rise to the valued apple variety *Blenheim Orange*. The original tree was found growing against a boundary wall at Blenheim Park, Oxfordshire and grew perhaps from a pip dropped by birds or from a labourer's discarded apple core. Once the seedling had developed into a mature tree, someone noticed that it produced tasty, distinctive apples. Cuttings from this one tree were grafted onto other apple rootstocks – and from these chance beginnings Blenheim Orange has spread to many different countries.

On other occasions, a variety's origin can be traced to deliberate attempts at controlled breeding: as when Thomas Knight, a Herefordshire squire, with the help of his daughter Frances, produced the first varieties of known parentage by cross-fertilising the *Golden Pippin* with other varieties in the early 1800s.[2]

The subject of apple varieties can be complex and confusing, as a bewildering profusion of varieties and synonyms has evolved over centuries across the length and breadth of the nation. Early on in the history of English apple-growing, things were probably simpler: *Old English Pearmain*, first recorded in 1204, seems to have been the main dessert apple in England well into the eighteenth century, being grown alongside its culinary counterpart *Costard*, which is why apple-sellers became known as costermongers. But dating old varieties with any confidence can be difficult; the Sussex variety known as Forge, for instance, was already widespread in northern parts of the county when Victorian cataloguers first officially described it in the mid-nineteenth century. Often a apple variety is dated simply to the year that it was "received" by the Royal Horticultural Society.

Over the centuries, a huge array of tasty English apples have been named and propagated. Many varieties probably arose in the eighteenth century, as wealthy gentry became keen to stock their walled gardens with high-status novelties. Exotic varieties also began to be imported into the UK, especially by Victorian entrepreneurs who brought new varieties from all over the world back to England for trials and development, adding to the wealth of home-grown varieties. There are now well over two thousand varieties of apple growing in the UK, and by 1966, when the Apple and Pear Development Council was begun, there were 150 named varieties of cider apple alone.[3] Below, we describe the varieties known to have Sussex origins which are presently recognised and map their distribution.

The naming of apples in Sussex

Apples often bear names that link them to a particular location, sometimes, perhaps, because they do especially well in the soil and microclimate that bred their original "mother" tree. But varieties may also be specifically bred or hybridised and then given a local name. One result is the profusion of fascinating names that we now encounter.

A good example of an apple variety seemingly adapted to its native soils is the *Wadhurst Pippin*, which grows well on the clay of its home village in the High Weald. Another variety named after its place of origin is *Hawkridge*, from Hawkridge Farm near Hellingly. The distinguished Victorian pomologist Robert Hogg wrote that "it is a very popular apple in East Sussex especially around Hailsham, Heathfield and those parts where it is met with in almost every garden".

And another is *Forge*, well known in north Sussex by 1851, when Hogg first described it, but probably much older, and which arose at either Forge Farm or near the old wealden ironworking sites around East Grinstead. Providing excellent eating around Christmas time, and also good for cooking and making cider, this apple was much favoured in cottage gardens.[4]

Some varieties named after places are misleading. *Alfriston*, a brisk cooking apple, was raised during the late eighteenth century by a Mr Shepherd of Uckfield and originally named Shepherds Pippin. Only later in 1819 was it renamed *Alfriston* after the apple was sent by a Mr Booker, who lived in the village, to the London Horticultural Society.[5] In a February 1890 edition of *The Garden* one writer refers to this variety as one of the best late-keeping varieties, a quality which we overlook today in our supermarket-dominated age.[6] The catalogue of West Dean calls it "the fishermen's apple, taken on voyages for its good keeping". Usually classed as a cooking apple, it was recommended for use as a dessert apple as late as March.

Other varieties have descriptive names, or are named after famous people. *Dr Hogg*, a large ribbed cooking apple raised by the head gardener at Leonardslee, Horsham, was named in honour of Robert Hogg who worked to rationalise and clarify the identity of the many different apple varieties grown in England in the 1870s.[7] Sussex varieties named after apple enthusiasts also include *Mannington's Pearmain*, named after John Mannington, the butcher noted above, who Hogg described as "a man of singular intelligence on many subjects and an enthusiastic pomologist."[8] The apple remained an entirely local fruit until specimens were sent to the RHS, after which it became more widely known and appreciated.

The well-known *Egremont Russet* was raised at Petworth House, the centre of the estate of the agrarian enthusiast, the third Lord Egremont, after his permanent move there in 1794 and before his death in 1837. The green skinned *Petworth Nonpareil* and the large conical *Bossom* were also raised at Petworth during this period. The last remaining tree of *Bossom* blew down during a gale in the village of Graffham and was only saved by having grafts taken by the Reverend Donald Johnson and Lady Caroline Egremont. The trees were then propagated by The National Fruit Collection at Brogdale and by the nursery of Frank P. Matthews. *Petworth Nonpareil* has not yet been found.[9]

Crawley Beauty and *Crawley Reinette* both originate from J. Cheal and Sons, the substantial and well-known nursery of 130 acres on the borders of Sussex and Surrey (see chapter 5). The latter, a dessert apple, was introduced by J Cheal & Sons in 1902. The *Crawley Beauty* was apparently discovered by Cheals in 1870 in a cottage garden in Lowfield Heath, now beneath Gatwick Airport, although another source offers the apple as being first grown at Martyr's Farm north of Crawley in Ifield parish.[10]

Other varieties have historical associations. The large, handsome, orange-red *Coronation*, streaked with dark red colouring, honours the Coronation of Edward VII in 1902. *Knobby Russet* describes exactly the appearance of the apple, with the underlying russet skin covered to greater or lesser extent with knobs, warts and lumps. *Tinsley Quince*, by contrast, is much more subtle, with its name describing the slight quince-like smell that gradually develops as the apple ripens.

The Sussex variety with the longest history in cultivation is *Golden Pippin* which has been grown since at least 1629. It is claimed that it arose at Parham Park but, as Joan

Morgan writes, its origin seems uncertain.[11] The very small yellow fruits have a brisk lemony taste and would have been cooked whole to make sweetmeats, jellies, tarts and cider.

The map shows the distribution of Sussex varieties for the first time. The distribution is clearly biased towards northern Sussex and the Weald, with clusters of named varieties around Crawley and Uckfield, and also around Petworth. By contrast, relatively few varieties hail from southern Sussex, especially the South Downs, where the soils of the open chalk downland are too dry and too thin for apple trees, except in some of the dry valleys where downwash has yielded deeper soils.

Natural variety

Sussex is famed for its geological diversity, with bands of distinctive soils stretching across the county, roughly east-west. So any traveller from Brighton to London, say, must pass through a series of strikingly different landscapes, some more suited to orchards than others.

Along much of the coast lies a strip of relatively warm and low-lying land with fertile soil. Much of this coastal plain, which runs from Brighton to Chichester and beyond, is a good place to grow fruit and vegetables, in the open or under glass. The Sussex apple variety *Aldwick Beauty* comes from this area, at Bognor Regis.

Just to the north lie the South Downs, a famous ridge of chalk, now a National Park. This windy upland sweeps from Eastbourne across the county to Winchester in Hampshire, and has thin and flinty soils not particularly suited to growing fruit. Yet even here, apple trees can thrive in sheltered coombes or river valleys where soils are deeper, and on gentle lower slopes of the downs.

Further north, at the foot of the downs, lie complex bands of different soils, derived from rocks including including greensand and gault clay. It was in the western end of the greensand, in the valley of the western Rother at Petworth, that the *Egremont Russet*, *Petworth Non Pareil*, and *Bossom* varieties originated. To the north again is the Low Weald which stretches from Hailsham west through Burgess Hill, widening around Billingshurst, and reaching north to Crawley and into Surrey and Kent. The Low Weald was once densely forested, and although the heavy clay makes much of the land unsuitable for arable farming, orchards can readily be established even here. At Hailsham, the *Hawkridge* variety was grown.

Finally, towards the borders of Kent and Surrey and overlapping into those counties, lies the High Weald, a large area of mixed sands and clays stretching from Horsham eastwards to Hastings. This region is rich in ancient woods and coppice and is mostly

The Sussex Apple Varieties

The map shows where each variety first came from; for some we have precise information, but not all. See the guide for more details

KENT
Forge
stead
Wadhurst Pippin
Ashdown Seedling
Crowborough
Wadhurst
Mayfield
Sussex Mother
Mannington's Pearmain
ckfield
Heathfield
Saltcote Pippin
Rye
Alfriston
Battle
Golden Bounty
Mareda
Hailsham
Hawkridge
Hastings
Bexhill
Polegate
Eastbourne Pippin
Eastbourne
WEALD

quite hilly, with flat-topped ridges cut by deep, steep-sided valleys known as ghylls. Apple orchards have long thrived particularly in this northern part of Sussex, which is home to *Wadhurst Pippin* and *Forge*.

A concise guide to Sussex apple varieties

Experts in apple identification learn to judge an apple by its shape and colour and flavour. Colours can range from yellow, red, green or russet, perhaps with subtle mixtures of hues, flecks and streaks, while characteristic shapes vary between flattish, rounded, conical, oblong, oval, angular and ribbed. The trees themselves vary in the way they grow, and some varieties grow more vigorously than others.

Apple Day, 2011: orchard expert Peter May guided visitors through the Stanmer apple-tree nursery and orchards now stocked with Sussex varieties

Another vital characteristic of an apple variety is the time of its flowering – roughly indicated by its F-number. It's good to know when a particular variety flowers because most apple varieties are not self-fertile, and need another variety that flowers at the same time to act as a source of compatible pollen. Honey bees, bumblebees and a host of other beneficial insects perform the vital transfer of pollen from one tree to another. In the following, the numbers F1 to F42 indicate when a particular

Apple shapes: one of the characteristics used by experts to identify apple varieties

variety will be in blossom and ready for pollination. F1 corresponds on average to the first of May, F31 to the end of May and F42 to June 11th , but these dates are only approximate and vary from year to year. To be able to successfully pollinate one another, the numbers of two varieties should coincide or overlap by four F-numbers before or three after.

Our summary of Sussex apple varieties draws on many sources and publications, but relies most heavily on the work of Joan Morgan and Alison Richards (see references at the end of this chapter). Additional observations have been made as part of the research into Sussex varieties by the Local Fruit Futures project.

Aldwick Beauty

Grown by Mrs DM Alford, Aldwick, Bognor Regis. Fruit is flushed red and becomes sweet by late August, but can become hard and bitter. The tree is moderately vigorous and the fruit ripens during August and early Sept. F17

Alfriston

Raised in the late-eighteenth century by Mr Shepherd of Uckfield and originally named Shepherd's Pippin. It was renamed Alfriston in 1819 by Mr Booker who lived in that village. It received an RHS Award of Merit in 1920. The medium to large apple is bright green or greenish yellow with conspicuous white dots (lenticels) on the skin of the fruit. The shape of the fruit is oblong with a flat base and a short thick stalk. Alfriston is quite a sharp cooking apple that cooks to a lightly flavoured puree and is very good baked. "It was much recommended and planted in the last century, when it proved to be one of the best culinary apples of the time. Now it is superseded by varieties more regular in outline and more handsome to the eye, though none is better for cooking," wrote HV Taylor in 1948. The tree is moderately vigorous and the fruit can be picked in early October. It stores well and can keep till April. F10 Biennial.

Ashdown Seedling

Grown by John Clarke who founded the orchards of the Ashdown & General Land Company during the 1930s. His son Rory Clarke remembers how his father grew it from the pip of a bright red McIntosh apple that he had eaten. The original tree grew in the kitchen garden at his home. It is an early apple that ripens in August with a light, sweet taste though Rory Clarke describes it as sometimes sour. A moderately vigorous tree. F8

Bossom

The last known tree growing in the garden of Miss Scott of Graffham blew down in the gale of 1986. On hearing of the damage to the tree the Rev. Donald Johnson and Caroline Egremont took graft material from the fallen tree and distributed it to the National Fruit Collection at Brogdale and later to the nurseries of Frank P. Matthews Ltd who propagated it for West Dean gardens. The apple is probably the variety described by Robert Hogg in 1851 and raised at Petworth House by Lord Egremont and Mr Slade. Hogg wrote: "Fruit large and conical, handsomely and regularly formed. Skin, pale greenish yellow, considerably covered with russet and occasionally marked with bright red next the sun. Flesh yellowish white, tender, crisp, juicy and sugary with a pleasant sub-acid flavour." The tree grows vigorously with upright growth and the fruit ripen in early October and can be stored until January. F12

Coronation

Raised by H C Princep at Buxted Park and introduced by nurseryman Pyne, Topsham, Devon. It was recorded in 1902 in the year of the coronation of Edward V11 and won an RHS Award of Merit in 1902. It is a large, handsome apple which can be variable in size and flat to almost conic in shape. The yellow skin is flushed orange-red and streaked with dark red. There is some russet at the base of the apple. Soft, coarse flesh with a sweet, light taste. Coronation is a vigorous growing tree with fruit that ripens in mid September and will store until November, though it becomes very soft. F13

Crawley Beauty

The apple was discovered growing in a garden in Tilgate near Crawley and introduced into cultivation by the local nurserymen J. Cheal. It was awarded an RHS Award of Merit in 1912. There is some confusion about its origin and it is identical to the French variety Nouvelle France (National Apple Register 1971). It is a very useful cooking variety that can be grown in colder, frost-prone areas because it flowers late and is disease resistant. The apple cooks to a lightly flavoured puree and can be eaten fresh after storing. It is a small to medium sized apple and is round and distinctively flattened at the base and apex. The colour of the skin is greenish-yellow, flushed brownish-red and with broken red stripes.

Flowering at the beginning of June the variety is self-fertile. A moderately vigorous, spreading tree which crops heavily and can be picked by mid October, storing until January. F29

❦ *Crawley Reinette*

Introduced by J Cheal & Sons, Crawley. It was recorded as a variety by the RHS in 1902. Quite sharp; becoming sweeter in October, with firm, cream flesh. The tree is moderately vigorous. The fruit is conic to straight-sided and the skin is yellow with an orange-red flush with some russetting. The flesh is greenish-white. Ripe in mid-October and can be stored until February. Reinette refers to the areas of russet on the apple, though it was originally intended to denote a fruit of quality (Saunders 2010). F15

❦ *Duck's Bill*

Introduced by Fred Streeter, Head Gardner at Petworth House in 1937. The apple is distinctively ribbed and flat sided which gives it its name. The skin is yellow, streaked pink with a red flush.There are russet dots. Quite rich tasting, sweet, but with plenty of acidity. The tree is medium vigour and very upright. The apples can be picked in early October and stored until December. There are other apples such as Winter Pearmain which have a similar appearance. F16

❦ *Dr Hogg*

Grown by Head Gardener Mr S. Ford of Leonardslee, near Horsham and thought to be a seedling from the variety Calville Blanc. Introduced about 1880 by W. Paul & Son, Herts and awarded an RHS First Class Certificate in 1878. A very large apple, conic and ribbed on body of the fruit. The skin is yellow, flushed pale red with slight stripes and russet patches. White flesh, sweet with slight acidity. It was named in honour of the Victorian pomologist Robert Hogg who wrote in his guide to the fruits and fruit trees of Britain: "....it melts perfectly, does not fall at all, is juicy, slightly acid, very rich and sugary with a pleasant aroma. It is a first rate baking apple" (Hogg 1851). A vigorous, upright growing tree. The fruit can be picked picked early September and stored until October. F9

Eastbourne Pippin

Grown by E.A. Lindley from Eastbourne 1930 and believed to be grown from the pip of a Newton Pippin. The apple is large, rectangular in shape and slightly ribbed at the base. The skin is golden yellow with a sweet, firm, creamy flesh that becomes soft and crumbly by October. The tree is moderately vigorous and can be heavy cropping. Fruit can be picked late August and stored until September or October. F7

Edmund Jupp

The apple was grown in the Horsham area and first recorded as a variety in 1862. It forms a small to medium sized fruit, flat-conical in shape and slightly ribbed at the base of the fruit. Yellow skin sometimes tinged orange with variable amounts of russetting. Flesh is creamy white. Robert Hogg (1851) describes it as an excellent culinary apple. The flavour is slightly acid with some sweetness. It forms a vigorous spreading tree. F13

Egremont Russet

As the name suggests it arose on the estate of Lord Egremont, Petworth. The head gardener Fred Streeter maintained that it was raised at Petworth by Mr Slater and Lord Egremont together with Petworth Nonpareil and Bossom. First recorded as a variety in 1872 by nurseryman J. Scott, Somerset, it received an RHS Award of Garden Merit in 1993. Distinctive flavour, often described as nutty; Morton Shand said it recalled the scent of crushed ferns. Almost smoky, tannic quality develops after keeping and flesh becomes drier. Popularity came early 20th century after nurseryman George Bunyard praised it as one of " richest late autumn fruits...pretty colour for dessert". The distinctive fruit is russet with a slight orange flush and a gold background colour. A medium sized apple that is flat round in shape. The stalk is very short and quite thin. Creamy flesh. A moderately vigorous tree that is resistant to scab and canker though prone to bitter pit. Pick the fruit in early October and store until December. F7

❦ *First and Last*

First described 1860. Hogg (1851) wrote that it is "...much grown in the northern part of the county about Horsham and sent to Brighton market". The shape of the apple is rectangular to conic with slight ribbing, with a lemon yellow skin, flushed red and streaked and blotched darker red with some russet. The flesh is a crisp, fine, greenish white with a slightly sweet flavour. It can be picked mid October and stored until April. At West Dean Gardens it is said to store for two years. F14

❦ *Forge*

The apple is also known as Sussex Forge and arose at either Forge Farm or near old iron forges around East Grinstead. Robert Hogg described it in 1851, but by then it was well known. Soon after picking it has a sharp tannic flavour but mellows to be almost rich, lightly aromatic. It cooks to soft, lemon coloured purée which tastes quite brisk and fruity. It also makes good sauce. The variety is known as the "Cottagers Apple" and was widely grown in north Sussex and Surrey in the nineteenth century. It is still found around East Grinstead. "A very old variety extensively cultivated throughout Sussex, where it is one of the most popular apples for dessert and cooking. This old variety still has many supporters," wrote H.V. Taylor in 1948. The fruit is a pale green-yellow with a light orange flush and a little red. Medium in size the apple is conical to round with a short thick stalk. The tree is a moderately vigorous grower and very resistant to scab, Martin Crawford reports. It can be picked late September and then stored until Sept-Dec or March if kept very cool. Hogg added: "it is a great and constant bearer". F15

❦ *Golden Bounty* *(no image available)*

Raised in about 1940 by A.C. Nash of Scutes Farm, Hastings. The parents of this variety are Cox's Orange Pippin crossed with probably Early Victoria. The fruits are crisp, sharply aromatic, like a brisk early Cox and can be very good. A moderately vigorous tree that crops heavily though the fruit is liable to crack. Pick during the middle of September and store until October. F13

Golden Pippin

The origin of the apple is uncertain as historically there were many similar apples with different names. It is claimed that it originally grew at Parham Park near Arundel. Golden Pippin was first recorded as a variety in 1629 by Parkinson though it was well known by late 1600s. It has an intense, brisk taste with a fruity lemon tang. When it is cooked it keeps its shape with a sweet, rich taste. The small fruit were poached whole to make sweetmeats, pippin, jelly, tarts and cider. Golden Pippin was widely grown in gardens and markets in the eighteenth century. In 1830 one commentator wrote: "This variety is too well known to need description as a small golden-coloured dessert apple of distinguished excellence" (quoted in Taylor). By the 1890s it was thought of as too small though it was used in the first apple breeding programmes. The fruit is gold coloured with many bold russet dots and some russet at the base. The apple is small in size and round-conical in shape with a short thick stalk and pale yellow flesh. The tree is moderately vigorous and resistant to scab (Crawford 2001). It can be picked in early October and stored until January. F11

Green Custard

Grown by the nurseryman J. Cheal & Sons of Crawley and received by the RHS in 1941. A large apple that is tall to conic in shape and ribbed on the body of the fruit. The skin is yellowish-green sometimes with faint orange flush and russet on the base. Greasy. Soft white to cream flesh that is slightly acid. The variety is very similar to Catshead which has many apples which are similar to it. The fruit can be picked in September and stored until November. F11

Hawkridge

Robert Hogg wrote that Hawkridge "is a very popular apple in East Sussex especially around Hailsham, Heathfield and those parts where it is met with in almost every orchard and garden" (Hogg 1851). The origin is Hawkridge Farm, Hellingly, near Hailsham. It is a medium sized apple, flat to almost rectangular in shape with prominent ribs on body. The skin is golden yellow with pale red stripes, flushed a dark bright crimson with darker stripes. The flavour is sweet, and sometimes described as balsamic. A vigorous upright tree. An early September apple, it can be stored until October. F12

❦ *Keed's Cottage*

Keed's Cottage comes from the Pulborough area and was recognised as a variety in 1947. A large apple, flat- rectangular to conic in shape and ribbed along its body. The skin is pale yellow, flushed orange-red and streaked red. It has a greasy feel. The flavour sweet and quite rich. The fruit is ripe in September and can be stored until December. F4

❦ *Knobby Russet*

Knobby Russet was first shown to the London Horticultural Society by Haslar Capron who came from Midhurst in 1820. The skin is covered in knobs or warts and is heavily russeted. A medium sized apple, flat truncate-conic in shape it has fine, firm rather dry flesh. A moderately vigorous upright spreading tree, it is resistant to scab though very susceptible to mildew (Crawford 2001). The apple is ripe in October and can be stored until December-March. F6

❦ *Mannington's Pearmain*

First grown about 1770 from cider pomace thrown under a hedge in the garden of Uckfield blacksmith, Mr Turley. His grandson, John Mannington, local butcher and keen fruitman, sent it to the London Horticultural Society in 1847. It has a rich, slightly aromatic taste with a nutty quality. It was well thought of by the Victorians. The apple is medium in size and conic to rectangular in shape with a fleshy protuberance at stalk. The skin is yellow flushed brownish red. Moderately vigorous in shape, it ripens in mid-October and can be stored through until February. Hogg wrote: "the tree does not attain a large size but is perfectly hardy and an early and excellent bearer. The London Horticultural Society 1847 pronounced it to be a dessert fruit of the highest excellence. Mr Mannington was a man of singular intelligence on many subjects and was an enthusiastic pomologist." (Hogg 1851). F13

Mareda

First grown in 1926 by A.C. Nash of Scutes Farm near Hastings, from a pip of the American apple Winesap. Handsomely flushed orange-red, and striped in bright red over yellow green. A large apple that is oblong-conical in shape, it has a distinctive sweet, winey or sweet melon taste, with firm flesh. The variety grows vigorously with biennial cropping. It can be picked in late October and stored until March. F12

Nanny

First grown in either West Sussex or Surrey and recorded as a variety in 1842. Hogg (1851) declared it to be: "a dessert apple of excellent quality and when in perfection a first rate fruit... This is met with about Havant and other districts of Sussex and on the borders of Hampshire." A large apple, flat-conic in shape it is slightly ribbed on the body of the fruit. The skin is pale yellow and covered with orange flush, dark red stripes and russet dots. Sweet with a rich flavour but soon becomes mealy. A vigorous, upright-growing tree. Pick early September and will store until October. F5

Petworth Non Pareil *(no image available)*

According to Reverend Donald Johnson the variety has not been seen in the past 20 years. It was probably bred by Mr Slater and the third Lord Egremont together with Egremont Russet and Bossom. First recorded as a variety in 1816, it is a medium sized apple, flat-conic in shape and not ribbed. The skin is green with a thin brown russet, firm, crisp, greenish flesh. The flavour is rich and slightly acidic with a picking season that is late to very late. Hogg (1851) judged it to have: "good flavour but not equal to that of the old Nonpareil. A dessert apple of second-rate quality."

❦ *Saltcote Pippin*

A variety which was cultivated by James Hoad of Rye from either a Ribston Pippin or Radford Beauty seedling. First recorded as a variety in 1918, it was awarded an RHS Award of Merit 1928. A large, handsome apple with a rich, aromatic flavour. It is sharper and not as intense as Ribston Pippin but has a similar, rather open texture. Becomes sweeter, milder, and more aromatic as it ripens. A medium-sized apple with a flat, rectangular to conic shape. The shape of the apple and the texture of the skin is often uneven. Pale to golden yellow in colour with a orange red to carmine flush with some russeting on the body of the apple. A moderately vigorous, spreading tree which crops well. Pick early October and then store until Nov-Jan. F11

❦ *Sussex Mother*

The variety was grown around Heathfield in the nineteenth century and described in 1884 by Robert Hogg. The apple is medium to large in size and tall to conic in shape with strong ribbing on the body of the apple. The skin is pale to golden yellow with an orange-red to carmine flush. The flesh is firm. It has a distinctive flavour, sweet, quite spicy with a hint of aniseed, though Hogg (1851) is less approving "an early dessert apple of no great merit". Still found in Sussex gardens. The tree is moderately vigorous with thin whippy growth. The fruit should be picked in early September and will store until October. Susceptible to mildew. F12

❦ *Tinsley Quince*

Received by the National Fruit Trials in 1942 from R. Fairman, Crawley. The tree is moderately vigorous and forms a spreading canopy of branches. The apple has a slight quince-like taste and smell when ripe. The flesh is sweet and firm. Fruit is greenish-yellow with a few russet dots and is slightly greasy. Flat to rectangular in shape and ribbed slightly on the body of the fruit. Pick early October and can be stored till November. F8

Wadhurst Pippin

Both Hogg and Bunyard write that the variety was grown early in the nineteenth century at Wadhurst. The fruit is large and variable in shape, tall to intermediate, rectangular to truncate or conic. Strongly ribbed on the body of the fruit. Skin yellow tinged green and flushed brownish-red, streaked crimson. The fruit has a savoury, spicy taste with quite a sharp, hard flesh. A very good cooking apple. A vigorous upright tree with fruit that can be picked early October and stored until February. F10

Varieties not listed in the National Apple Register

Lady Hopetown

Entered National Fruit trials in 1950, sent by W Barnes of Bexhill-on-Sea. The fruit is greenish yellow with a crisp sharp fruity flavour and can be picked in mid October. Can be stored until March and on to May if kept very cool. F11

Queen Mary *(no image available)*

An article in the *Gardeners Chronicle* of November 1, 1919 describes how this apple was introduced by E.J. Parsons of Sussex and then selected by the RHS Fruit and Vegetable Committee as the "best seedling apple submitted to them" with "flavour reminiscent of Cox's Orange Pippin". The parents of the fruit are *James Grieve* x *William Crump*. The apple has a yellow skin streaked and shaded with crimson scarlet. The variety is currently grown by Graham Bedding, Apple Revival, Cavendish, Suffolk. The National Fruit Collection at Brogdale prefer to call this variety Duchess of Oldenburg. F4

Stanmer Pippin

The tree was grown from the pip of a "supermarket apple" by Anne Markwick at Stanmer Park during the 1960s and raised by Plumpton College Horticultural Training Centre, Stanmer Park, Brighton and Hove. The fruit is medium in size with a round conical shape. The skin is flushed orange-red with broken red stripes on a yellow green background with some russeting. There are prominent white dots on the skin (lenticels) and it is slightly greasy. The flesh crisp and fine textured with greenish tinge, juicy and sweet with a rich flavour. It forms an upright, spreading tree that is moderately vigorous. The fruit hangs on the tree and can be picked in mid-October. F12?

Sussex Apple List References

Bedding, G. *Apple Revival*, Cavendish: Suffolk.
Bultitude, J. *Apples. A Guide to the identification of international varieties.* Seattle: University of Washington Press 1983.
Crawford, M. *Directory of Apple Cultivars*. Dartington, Devon: Agroforestry Research Trust 2001.
Egremont, Max. *A Week in the Country*, article on Petworth Apples. *Country Life*, January 1991.
Hogg, R. A *Guide to the Fruits and Fruit Trees of Great Britain*. Fifth ed. 1884 Journal of Horticultural Office (first published 1851).
Johnson, Rev D. *pers comm.*
Morgan, J. and Richards, A. *The New Book of Apples*. London: Ebury Press 1993.
Miller, Jim, *pers. comm*, Plumpton College Horticultural Training Centre at Stanmer Park, Brighton and Hove.
Saunders, R. *The Apple Book*. Phaidon Press 1986
Smith MWG. *National Apple Register of the United Kingdom.*1971.
Taylor, HV, *The Apples of England*. Revised third edition, London: Crosby Lockwood and Son 1948.
West Dean Gardens, nr Chichester, West Sussex, PO18 0QZ.

For more information on choosing apple trees, please see chapter 12.

Endnotes for Chapter 2

1 Liberty Hyde Bailey, *The apple tree* New York: Macmillan, 1922, 68.
2 J. Morgan and A. Richards, *The new book of apples* London: Ebury Press 2002, 78.
3 http://yourdemocracy.newstatesman.com/parliament/apple-and-pear-development-council-order-1966/HAN9101858 (accessed 26 May 2010)
4 Anon, 'Fruit at Crawley' *The Garden* 1 Nov. 1890, 419.
5 Morgan and Richards, 184.
6 'E.H.'. 'Apple Alfriston' *The Garden* 1 Feb 1890, 93.
7 See chapter 4 in the present volume.
8 Hogg, 1851. See Morgan and Richards, 239.
9 Donald Johnson, pers. comm. 2008.
10 W. E. Cheal, *A Family Business* (copy in Crawley Ref. Library), 54.
11 C. Macintosh, *The orchard: including the management of wall and standard fruit trees* London: W.S. Orr and Co. 1839, 21.

3. Beginnings

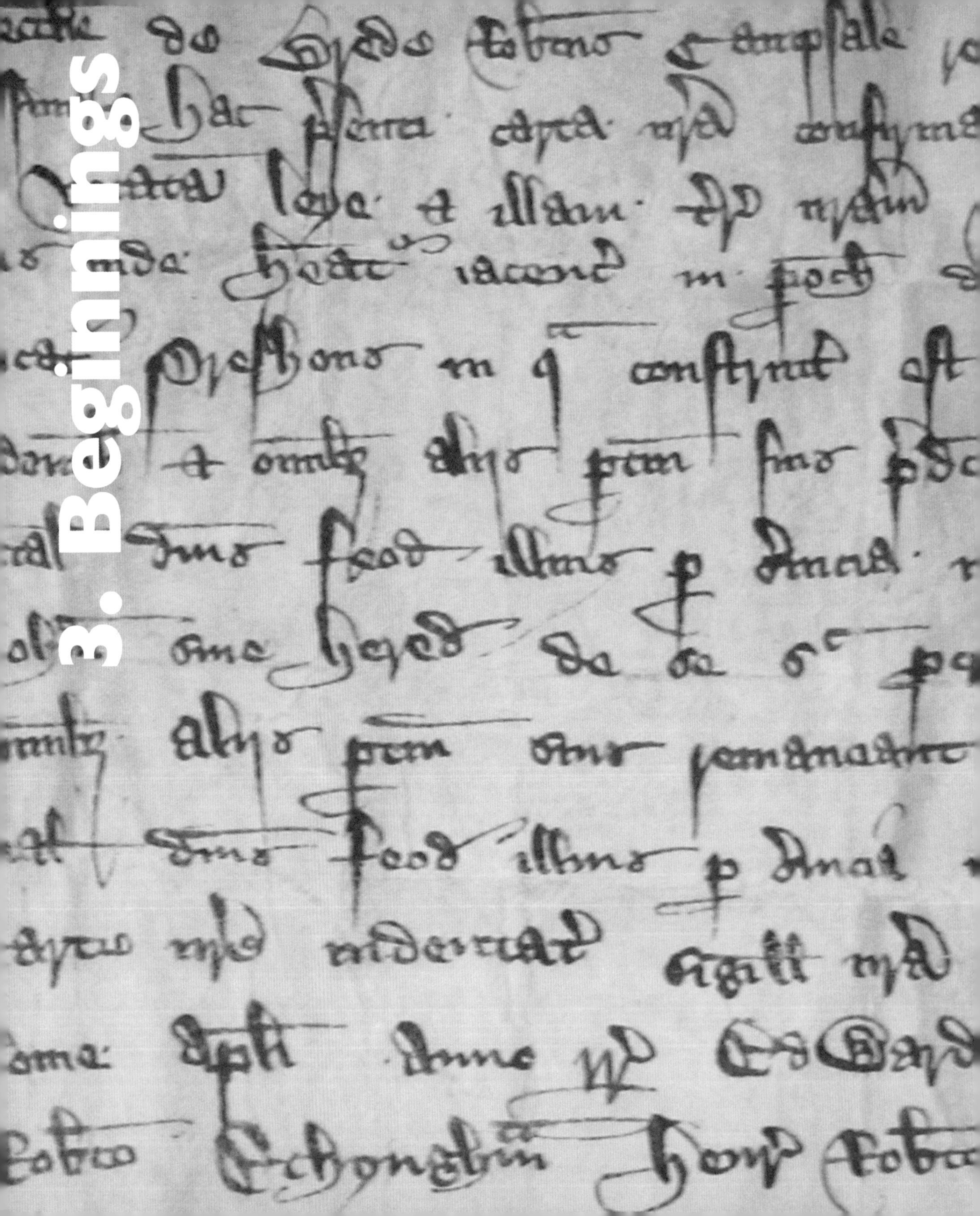

3

Apples in Sussex: beginnings

The Sussex apple story begins far from southeast England, in the remote mountains of central Asia. All of today's varieties of orchard apples – which share the Latin name *Malus domestica* – are now known to be direct descendants of the wild apples *Malus sieversii* that still grow in the natural forests of southern Kazakhstan. DNA tests by Barrie Juniper of Oxford University and other research teams around the world have largely solved the mystery long shrouding the apple's origins.

The modern apple began to emerge in the foothills and valleys of the Tien Shan mountain range, thousands of years ago. Wild mammals with a sweet tooth – including bears, deer, wild pigs, wild horses and donkeys – selected the tastier, more nutritious apples and spread the pips via their dung, thereby driving the evolution of more palatable fruit. Once people and domesticated horses arrived on the scene, the

Facing page: valuable medieval asset - a 1373 land valuation document lists land at Northiam on which there were two buildings 'one called Preshous in which stands an apple-mill'

biggest and best wild apples travelled westwards stowed in saddlebags, as generations of traders journeyed along long routes later known as the silk roads.

The home of the modern apple – the apple Eden – lies near Kazakhstan's former capital – Almaty, a city with over a million people, and once known as Alma-Ata, variously interpreted as meaning "place of apples" or "Father of Apples".[1] Here, the remnants of once-huge forests on the slopes of the Tien Shan mountains still harbour

Apple Eden, the Tien Shan of Kazakhstan: the foothills of these relatively young mountains are covered with forests of wild apple trees - ancestors of our domesticated varieties

a vast genetic diversity of wild apple trees. The environmentalist and author Roger Deakin, writing in 2008 after travelling there, marvelled at the wild apple woods that still cover the mountain slopes and valleys, gleaming white with apple blossom in May. Bee-keepers move their hives up and down the hills with the seasons, in old Russian lorries.[2]

It may have taken as long as 6,000 years for the wild apples to reach the Atlantic shores of western Europe, where they thrived in regions that were too cold for grapevines. In these new surroundings, these resourceful trees continued to adapt, producing new genetic mixes drawing on the species' rich inheritance from its central Asia homelands.

Grafting breakthrough

No one knows when apple trees were first propagated by grafting – through the binding of a cutting from a favoured tree onto the stump of a sapling – but the techniques may well have been discovered by early farmers. Both the Chinese and the Romans regarded grafting as a very ancient art, and it was a well-known practice in the Mediterranean by the fifth century BC.[3] Grafting was a breakthrough in orchard technology because it enabled people to easily propagate the trees they particularly liked in great number. Grafting provided a way round the difficulty that most apple varieties do not root easily from cuttings, and also don't breed true from seed – each apple pip can potentially produce a new kind of apple that may bear little resemblance to its parents.

Yet the extraordinary genetic diversity packed into every apple core has stood the apple in good stead on its journeys round the globe. By planting apple seeds and picking out the most promising young trees that result, growers can foster new varieties that may turn out to be much better suited to the local soil, pests, climate and day length. This approach proved to be a godsend for the first colonists to the New World from England in the seventeenth century, for instance, when the (grafted) apple trees they imported failed to thrive in America. Sensibly, the colonists also planted apple seeds, and among the so-called "seedling" trees that sprang up, a few had just the right genetic mix to prosper in the new land, and were fruitful and hardy. Such prized specimens were then readily multiplied by taking cuttings and grafting them onto the lower stem and roots – the so-called "rootstock" – of local wild crab apples. Every new young grafted tree was genetically identical to that of the seedling tree that had germinated and thrived in the new land.

Alluring apples

Such propagation techniques enabled apples to spread throughout the ancient world. References to apples in classical literature abound, and a vast mythology developed linking golden apples with temptation, courtship and marriage. In the Biblical story of Adam and Eve, the forbidden fruit to which Eve succumbed is typically portrayed as an apple – perhaps reflecting the influence of Greek mythology on Renaissance painters depicting the story. Throughout history, depictions of the Garden of Paradise include apple trees, and an association with rebirth, youth and beauty and magical powers seems also to have developed. Stories often focus on the Hesperides, the nymphs living in a beautiful garden bearing the tree with the golden apples – a wedding present from Gaia to Hera on her marriage to Zeus. This garden was guarded by Ladon, a dragon with a hundred heads, and only Heracles, through trickery, succeeded in obtaining some of the apples, thereby completing the eleventh of his Twelve Labours.[4]

The Roman Goddess Pomona, in Vertumnus and Pomona by Hendrik Goltzius, 1613. Pomona was the goddess of fruit and fruitfulness and wife of Vertumnus, the god of orchards, who persuaded her to marry him after having tricked his way into her garden disguised as an old woman

With the rise of the Roman empire, apple production moved further north and west. Italy, according to Horace in the first century BC, had vast areas of orchard. The larger, sweet apple, referred to as "urbaniores", was held in high esteem and preferred to the Germanic tribes' wild apples.[5] The Roman Goddess Pomona, whose insignia was the pruning sickle, oversaw fruit trees, particularly their blossoming.[6]

Many cultivated varieties were recognised by the Romans, some for their medicinal or even aphrodisiac properties. Retreating from the noise and wearisome politics of the city, Roman elites could relax in their villas surrounded by their own orchards, offering not only fresh fruit but appealing scenery.

Apples in Sussex: the first millennium

The crab apple (*Malus sylvestris*) grows wild in Britain, and archaeological excavations suggest that from the Neolithic period onwards, England's prehistoric inhabitants gathered its small, hard fruits to eat, cook or dry for storage and perhaps even fermented them to make an alcoholic drink. Wild-food expert Ray Mears says modern crab apples vary greatly in their sweetness, and he imagines that our ancestors foraging through the ancient wild woods would have learned which trees were best. In his 2007 book with Gordon Hillman, *Wild Food*, he recommends roasting or baking the fruits, best gathered after the first frosts have sweetened them.

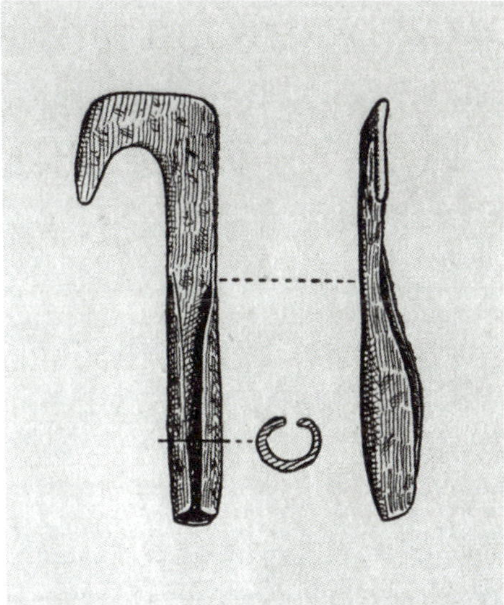

Ancient traditions such as wassailing apple trees on the eve of Twelfth Night or bobbing for apples in containers full of water at Hallowe'en may date from this pre-Roman period along with many myths linked to apples from around western Europe. In ancient Cornish, a Celtic language, apple is *aval*, and the mythical burial place of King Arthur was the Isle of Avalon, or the Isle of Apples.

Domesticated, grafted apple trees probably arrived in Britain with the Romans, who landed on the southern coast of England in 43AD – perhaps, some scholars now believe, in the Chichester/ Solent area. Over more than four centuries of Roman occupation, until its collapse in the fifth century, these powerful incomers established rich farms – the villas – as well as forts, towns and smaller settlements, and imported their own apple varieties, the domesticated descendents of the apple forests of Kazakhstan. To establish orchards, the invaders may well have grafted their own sweet apples onto stumps of the wild crab apples. It is said that one present-day apple variety, *Court Pendu Plat*, or possibly *Court Pendu Gris*, was known to the Romans.[7] Pruning knives and other possible signs of pomological activity have turned up in the archaeological record, and what is believed to be a Roman pruning hook has been found at West Blatchington (Hove).[8]

Cutting edge: found at West Blatchington, Hove, some years ago, this tool might have been used by Roman gardeners to prune apple trees.

The Roman villa at Fishbourne, near Chichester, gives stronger evidence of an orcharding tradition in Romano-British Sussex. On the eastern side of the reconstructed formal garden at the great Roman palace, just south of Chichester, archaeologists have found alternating tree-pits and postholes, suggesting an original arrangement of trees or shrubs planted against a trellis. Indeed, Pliny the Elder referred to espalier fruit trees planted against a trellis in his own Roman gardens, where they "added an air of rustic simplicity to an otherwise formal setting". Although the fruit trees at the Fishbourne villa might not have been apples, espalier apple trees (bearing modern Cox apples) have now been planted on the site. Wild crab apples are also grown at Fishbourne.[9]

Echoes of the past: at Fishbourne Roman Palace, espalier Cox apple trees have been planted where fruit trees grew in Roman times

Apples in medieval Sussex

With the fifth-century fall of Rome, the Roman occupation of England came to an end, yet orchards and the knowledge of how to cultivate them may have persisted through a period little documented in the written record. Soon there were new incomers – this time, in Sussex, particularly from Saxony in what is now Germany, hence "Sussex", from South Saxons. Both archaeological and linguistic evidence testifies to the enduring importance of apples and orchards in Sussex through the subsequent Anglo-Saxon era, which lasted for more than 500 years and ended only with the arrival of the Normans in 1066.

The Saxons evidently made good use of the crab apples growing wild in the landscape –as suggested by the size and shape of seeds discovered during excavations of an early Saxon settlement at Bishopstone, near modern Seaford, for instance.[10] But it also seems likely that in many areas the Saxons continued to harvest fruit that had been cultivated in the Roman period, and re-established the art of orcharding themselves.[11]

The best evidence for the continuing cultivation of apples comes from place-names. Sussex retains several place-names, such as Apuldram parish near Chichester, containing the Saxon word *apulder*, apple tree, with *-hamm* indicating a place hemmed in by winter or marsh, in an area with apparently apple-friendly soils. "To this day the land [of Apuldram], a deep loam with a clay or brick-earth subsoil, is admirable apple-growing land," wrote W.D. Peckham in the late 1920s, to the compilers of *The Place-Names of Sussex*.

Applesham Farm near Lancing also includes the name, although the "apple" element might here be a personal name. There is also Crabtree near Cowfold, and Adder Wells in the parish of Firle, north of the present A27, meaning "apple tree wells" (there are springs near here).[12] During the Anglo-Saxon era, apples were clearly important enough in some areas of Sussex to be meaningful signifiers of place.

The word "orchard" itself also has Anglo-Saxon roots, from the Old English *ortgeard*, a garden or orchard, *orcerdlēh* with a similar meaning, or more precisely *apuldortūn*, an apple orchard.[13]

With Augustine, who travelled from Rome in 597AD with a mission to convert the inhabitants of England to the Christian faith, came the tradition of monastic communities. But it was not until nearly a century later, around 681, that the new religion put down roots in Sussex – famously the last English kingdom to be converted. These early Christian establishments fostered the tradition of apple-growing, and by the tenth century, the Abbot of Ely was noted for being "skilled in planting gardens and orchards", while an eleventh-century Anglo-Saxon document lists the preparation of orchards and the grafting of new trees as the work of the bailiff.[14] By this time too, the Old English *ortgeard* had become *orcead* and hence to the Middle English *orchard*, but with recognition of the second element *-yard*, an area of land.[15]

The arrival of the Normans

Apple cultivation in England received a significant boost after the Norman conquest of 1066. These invaders from northern France may even have introduced the art of cider making – which had long thrived in Normandy – or at least refined existing techniques, for the design of the oldest cider press documented in England resembles the equipment traditionally used for making French wine and olive oil.

At Lewes, the chief house of the English Cluniac order was established soon after the conquest, with the great priory church of St Pancras the largest and most spectacular in Sussex until it was destroyed in the 1540s. The monks tended walled grounds stretching over 32½ acres, enclosing a paradise (park), a garden and an apple orchard, with plums and pears.[16] The twelfth-century expansion of the Cistercian order of monks, with their emphasis upon the cultivation of their own lands, gave another boost to apple-growing – in Sussex, the order established a monastic community at Robertsbridge.

Orchards and well-stocked gardens were also features of many town houses by the end of the eleventh century. Houses in early towns such as Chichester, Arundel, Steyning, Lewes, Hastings and later Winchelsea were typically built on an elongated plot of land (the burgage plot), where orchards and vegetables were cultivated and livestock kept.

In the earliest known reference to English-grown apples, dating from 1290, the apples were called Pearmains, and might have been used largely for making cider. In medieval England, apples were rarely consumed as fresh fruit, but were typically cooked with spices or used as verjuice – a sharp, slightly fermented apple vinegar – or drunk as cider, used medicinally, or simply admired as decorative features of ornamental trees in the pleasure grounds and gardens of the wealthy.

Throughout the thirteenth and fourteenth centuries, cider was widely drunk and traded in Sussex; records show that cider, as well as wine, was being imported from Normandy through Winchelsea by 1270. Supplying vitamin C and vitamin B_{12}, the unpasteurised drink was certainly beneficial, in contrast to the often-contaminated drinking water. Special cider apple varieties were developing in the west of England, but Sussex cider-makers fermented blends of culinary and dessert varieties, making an alcoholic brew enjoyed across the social spectrum.

Wages or rents were frequently paid, in part, in cider. A document from Durford Abbey, Rogate, recorded that cider was to be given to John Cook on 27 March 1323 as part of a "corrody" – a grant of land or a plot plus sustenance in return for a gift to the Abbey, in this case £40 for repair of the abbey church.[17] The manufacture of cider was also a way of raising income: Battle Abbey was selling cider from its estates by the early thirteenth century, and the cellarer's accounts noted that in 1369-70, 55 shillings was made from the sale of 3 "tuns" of cider from the garden, "12 tuns having been deducted as expenses, and 20s, for the purchase of barrels and the collecting of apples".[18]

In Sussex tax returns, both ecclesiastical and lay, cider often featured as payment. Taxes owed to the church, the "tithes", were regularly paid in cider. These documents show that in northwest Sussex, the parishes of Harting, Sutton, Easebourne and Kirdford produced significant amounts of cider: one calculation, based on the tithes taken before an extra tax was levied by Edward III in 1342, estimated that there were 200 acres of apple orchards at Kirdford, while Harting may have had as much as 500 acres, and Sutton 150 acres – not inconsiderable amounts.[19]

In 1332 so common were orchards, and indeed the production of cider, that as many as 74 out of 80 western Sussex parishes paid some of their church tithes in cider.[20] Tithes were payments to the church for the support of a priest for the community, and during the 1300s tithes of cider were being collected everywhere from West Angmering and Barpham (a medieval deserted settlement north of Angmering)[21] to Petworth, Sutton, Eridge, Duncton and East Preston.[22] Some 20 tuns of cider were sent from Field Place Goring, to Shoreham and on to Calais in 1349, and cider was still being made here in the 1520s and 1530s.[23] At Poling the home farm of Poling Manor had an apple mill for cider in the 1360s.[24]

Entries for cider abound in Petworth ministers' accounts (1347-53), such as the £4 4s received for 7 casks of cider, underlining cider's important role in the local economy.[25] In 1347, apples were being grown on some scale on the Petworth estate: 251 young apple trees at 6s 8d the hundred and 24 pears at 4s 8d were purchased for the "lord's great garden", along with three young trees for a garden within the Great Park.[26] Payments were also made for repairing and hooping cider casks, and for supplying 1lb of grease for the cider press and 2½ gallons of tar, applied to the trees to protect them from rabbits. A short account by Richard Goslyn, gardener, for 1375-6 mentioned the sale of "Genetz" pears and "Wardonns".[27] Across the country, manorial lords produced vegetables, fruit, herbs, honey and beeswax for household consumption, and sent surpluses to local markets.[28]

Surplus cider generated income for farms and estates. At North Stoke, near Amberley, the manor home farm was selling cider in the 1350s (and still had an apple orchard much later, in the 1520s).[29] At Climping the Cudlow home farm sold cider in the

1360s.[30] Cider-making equipment featured in land valuations: a grant made in 1373 listed "Land called Leye and land called Lucasesmarlyng (20 acres) in Northiam with the two houses built thereon, one called Preshous in which stands an apple-mill". Another grant of land contained an "apple-mill on the land, all together at Swiftsden in Etchingham" in 1402.[31] In the 1380s "men from the galleys" were said to have attacked Appledore and taken, among other things, two pipes of cider.[32] Appledore, just across the border in Kent and on the edge of what is now Romney Marsh, is named after the Saxon *apuldre* for "apple-tree", and the parish was thus a "place at the apple-tree".

Further evidence that trading in fruit was growing in importance in medieval England is provided by the founding in London of the Worshipful Company of Fruiterers. One of the oldest City companies, dating its official grant of ordinances to 1463, the Worshipful Company of Fruiterers regulated all aspects of the work of fruit traders and also attempted to protect their interests. Somewhat later, but in perhaps the earliest instance of what will become a familiar theme in our fruit-growing history, the company petitioned members of Parliament in 1622 to prevent the sale of imported fruit which was threatening the viability of orchards, referred to as "the plantation", which though "soe beneficiall both to poore and ritch is like to fall to great decay," as A.W. Gould recorded in his 1912 *A History of the Worshipful Company of Fruiters* (p.xii).

Tempting work: Adam and Eve and the apple tree form the central motif of the Worshipful Company of Fruiterers

Orchard work was sometimes stipulated in medieval custumals, the recitations of tenants' obligations to their lord. In 1374 John Loteby, the holder of a small cottage belonging to the Bishop of Chichester's manor of Stretham, near Henfield, was obliged, among many other duties, to "plant potherbs for two half days before or after dinner, and shall gather apples for two half days, before or after dinner".[33]

One final example from medieval Sussex comes from the accounts for Sir William Scott's Manor of Mote in the parish of Iden, but covering also parts of Peasmarsh. The entry for the year 1472-73 described the expenses involved in planting a new orchard:

"John Gardener, hired for making the lord's garden beside the old hall within the moat, and for supporting the vines there with pales 5d.
Carrying 150 apple trees, bought at Hawkhurst by Vincent Finch, to Newenden 1s 4d.
Bringing them to the Mote by water, with the expenses of Thomas Longbregge and Edward Parrok working on them 1s 3d
Planting the trees 3d."

And for 1474-75:

"Paid Henry Asshmeston and William Amyot, hired for planting apple-trees sent to this manor by Vincent Fynche in the lord's new garden 1s 2d."

Two years on (1477-78), the apple trees were being pruned and fertilised:
"William Pellond, hired for pruning and cutting the tap-roots (*ramis aquaticis*) of the apple trees in the lord's garden in winter, and for removing the earth from their roots and replacing it with new earth, working for eight days with the lord's servants 2s 1½d."

Well over a century later, a sketch map of 1633 depicted an orchard on what could be the same site, with trees set out in rows next to the manor house within the moat.[34]

These earliest glimpses of apple cultivation in Sussex suggest a vigorous early tradition of apple-growing in the county. The practice of cultivating domestic apples introduced

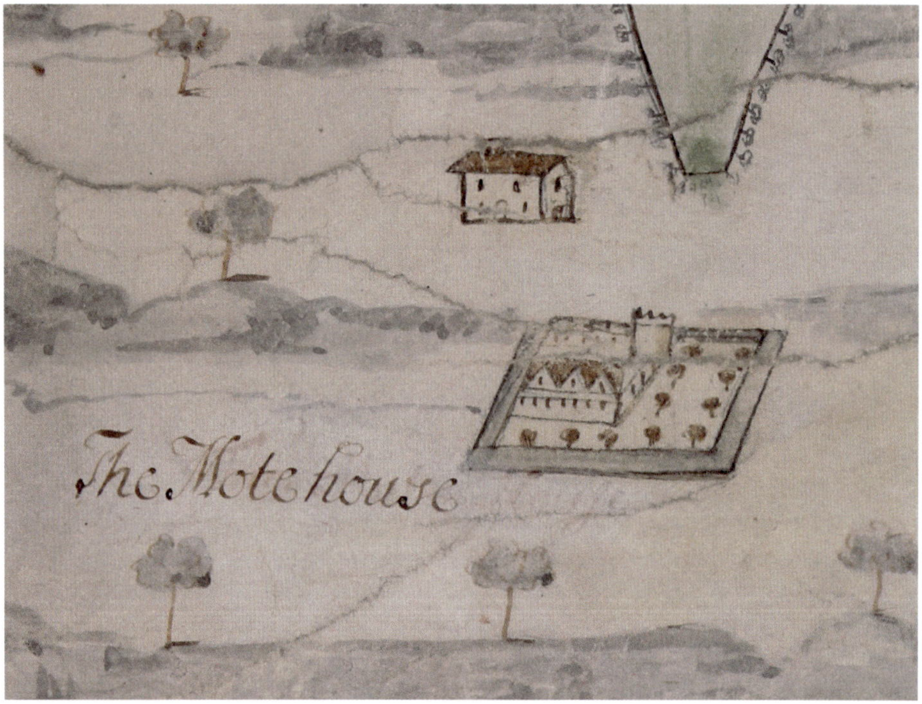

Good investment: in the 1470s, Sir William Scott spent more than six shillings buying apple trees and hiring gardeners to plant and prune them. Long after he was gone, his orchard was probably still there – as suggested by this 1633 map of Mote House in the Sussex parish of Iden

by the Romans evidently flourished through the Saxon period, only to be boosted by the arrival of the Normans – as an obsolete Sussex phrase for an orchard with a French flavour – "apple-terre" – suggests.[35] As surviving documents testify, apple-growing, especially for cider making, became widespread across Sussex in the medieval period as the orchard became an integral feature of both town and country.

Endnotes for Chapter 3

1 J. Russell, *Man-made Eden: historic orchards in Somerset and Gloucestershire* Bristol: Redcliffe Press 2007, 19-20; J. Morgan and A. Richards, *The new book of apples* London: Ebury Press (rev edtn 2002), 9-23.

2 R. Deakin, *Wildwood: a journey through trees* London: Penguin 2008, 283-346; http://news.bbc.co.uk/1/hi/sci/tech/8036785.stm (BBC Science page, 'Wild fruit trees face extinction' 7 May 2009); B. Juniper and D. Mabberley, *The story of the apple* London: The Timber Press 2006, 17-80.

3 K. Mudge, J. Janick, S. Scofield and E. Goldshmidt, 'A history of grafting' *Horticultural Reviews* 35 (2009), 449-59.

4 The classical apples of the Hesperides may in fact have been citrus fruits, or quinces (Juniper and Mabberley, *The story of the apple*, 153).

5 Juniper and Mabberley, *The story of the apple*, 127.

6 Morgan and Richards, op cit., 16.

7 Morgan and Richards, op cit., 202.

8 N. Norris and G. Burstow, 'A prehistoric and Romano-British site at West Blatchington, Hove' *Sussex Archaeological Collections* 90 (1951-2), 239, and plate xii, 3.

9 http://www.fishbourneromanpalace.com/about/the-gardens. We are grateful to the staff at Fishbourne Roman Palace for their assistance with information about the espalier trees there.

10 G. Thomas, *The later Anglo-Saxon settlement at Bishopstone: a downland manor in the making* York: Council for British Archaeology Research Report 163 (2010), 167-70.

11 H.P.R Finberg (ed), *The agrarian history of England and Wales I* (ii) A.D.43-1042 Cambridge: Cambridge University Press 1972, 422.

12 We are grateful to Professor Richard Coates, University of the West of England, for expert guidance on Sussex place names. He suggests that both Appledram and Applesham denote Anglo-Saxon apple locations.

13 J.R. Clark Hall, *A concise Anglo-Saxon dictionary* Cambridge: Cambridge University Press 1960; P. Dallas, *Orchards in the Norfolk landscape: historic evidence of their management, contents and distribution* (2010) [http://www.norfolkbiodiversity.org/pdf/reportsandpublications/Orchards – accessed 9 March 2011]

14 H.P.R. Finberg (ed), *The agrarian history of England and Wales I* (ii) A.D.43-1042 Cambridge: Cambridge University Press 1972, 512-16; Morgan and Richards, op. cit., 26.

15 OED, 'Orchard'.

16 http://www.lewespriory.org.uk/aerial_view

17 J H Stevenson *The Durford cartulary* (Sussex Record Society 90, 2006), 83.

18 T. McLean, *Medieval English Gardens* London: Collins 1981, 12; French, op. cit., 12.

19 E.M. Yates, 'Medieval assessments in north-west Sussex' *Transactions Institute of British Geographers* 20 (1954), 75-92. The help of Pam Combes, whose work on the Sussex Inquisitiones nonarum is just beginning, is gratefully acknowledged.

20 Morgan and Richards, 138.

21 *Victoria county history Sussex* 5 (ii) 2009, 47.

22 *Victoria county history Sussex* 5 (ii) 2009, 257.

23 *Victoria county history Sussex* 5 (ii) 2009, 116.

24 *Victoria county history Sussex* 5 (ii) 2009, 243.

25 L.F. Salzman, 'Early taxation in Sussex, Part II', *Sx Arch Coll.* 99 (1961), 1-19.

26 West Sussex Record Office Add Ms 12238.

27 L.F. Salzman, *Ministers' accounts of the Manor of Petworth* 1347-1353 Sussex Record Society 55 Lewes, 1955, xxx-xxxi, 3-7.

28 B.M.S. Campbell and K. Bartley, *England on the eve of the Black Death: an atlas of lay lordship, land and wealth, 1300-49* Manchester: Manchester University Press 2006, 105-9.

29 Arundel Castle Mss M314, m.5.

30 *VCH Sussex* 5(1) 1997, 117-27.

31 East Sussex Record Office (ESRO) FRE 6909 21 Dec 1373; AMS 4666/14 18 Apr 1402.

32 E. Miller (ed), *The Agrarian History of England and Wales Vol. 3, 1350-1500* (1991), 125.

33 W.D. Peckham, *Thirteen custumals of the Sussex manors of the Bishop of Chichester* Lewes: Sussex Record Society 31 (1925), 114.

34 M. Gardiner and C. Whittick (eds), *Accounts and records of the Manor of Mote in Iden* Lewes: Sussex Record Society vol. 92, 2011, xxxvi, xlviii-l and 71, 78, 106.

35 W. Holloway (of Rye), *A general dictionary of provincialisms* Lewes: Baxter and Sons 1838, 4.

4. 1500–1700

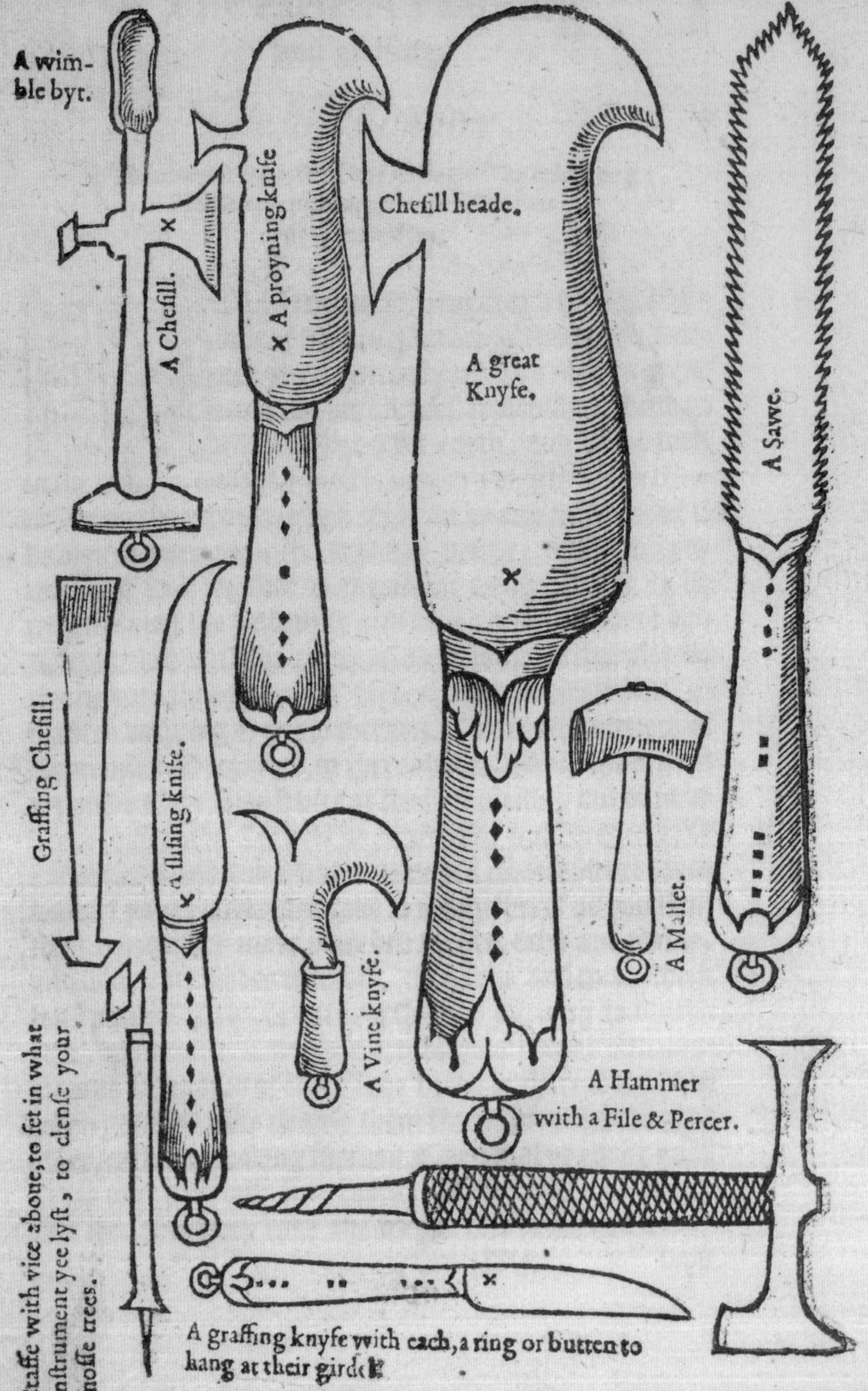

A wimble byt.
A Chesill.
A proyning knife
Chesill heade.
A great Knyfe.
A Sawe.
Graffing Chesill.
A slising knife.
A Vine knyfe.
A Mallet.
A Hammer with a File & Percer.
Staffe with vice above, to set in what instrument yee lyst, to clense your mosse trees.
A graffing knyfe with each, a ring or butten to hang at their girdle.

4

Apples in Sussex: 1500-1700

"Thy breath is like the steame of apple-pyes."
Robert Greene (1590), Arcadia

By the sixteenth century, knowledge of orcharding techniques began to spread, as printed books became more widely available. John Fitzherbert's *The Boke of Husbandry* (1523), with its descriptions of grafting procedures, was one example of the burgeoning interest in apple-growing across Europe.

A dramatic shift in orcharding traditions in England came during the 1530s and 1540s, as Henry VIII broke up the monastic landholdings that had dominated much of the Sussex landscape. The church lands were taken over by the king, and then

Many cottagers and small farmers kept a few sheep, hens or pigs in their orchards, making the most of the shade, grassland and fallen fruit they provided

A conjectural aerial view, from the west, of The Priory of St Pancras, Lewes, in the years immediately prior to its dissolution in 1538. Orchards were prized features of such ecclesiastical establishments

disposed of to favoured subjects – gentry, lawyers and government officials. The former ecclesiastical lands, with their orchards, were now in private hands. Battle Abbey and Lewes Priory, for example, were suppressed in 1538, with the 22 manors belonging to Battle Abbey going to Sir Anthony Browne.

The new gentry, suddenly major landowners, began to create their own pleasure gardens bedecked with decorative fruit trees, and to show off their knowledge of classical authorities and their writings on orchards.[1] Fresh fruit at the end of the meal, as in Roman times, became a status symbol. All the while, apples continued to be a valued source of cider in the Weald of Sussex, which, after all, was the perfect alcoholic brew for a landscape where the barley needed for beer and ale was difficult to grow. Writing in 1576, one commentator alluded to the "Wealdish, or woody places, where of late daies they bred much pomage, or cider for want of barley… ."[2]

Rather tenuously credited by many with the introduction of the "pippin" to England – perhaps a particular variety of good eating apples – was Leonard Mascall (d. 1589) from a family well-known in Sussex. Living in Plumpton Place, near Lewes, Leonard was described thus by Fuller in his *Worthies of Sussex*:

> Leonard Mascall, of Plumsted [sic] in this county, being much delighted in Gardening, man's Original vocation, was the first who brought over into England, from beyond the seas, Carps and Pippins; the one, well-cook'd, delicious, the other cordial and restorative.

According to Mark Anthony Lower in his own *The Worthies of Sussex* (1865) Mascall planted his pippins "in his orchard, hard by".[3]

Confusingly, in 1569 another Leonard Mascall, Buckinghamshire-based and unrelated to the Sussex Mascall, published the *Book of the Arte of and Manner howe to plant and graffe all sortes of trees, howe to set stones, and sowe Pepines to make wylde trees to graffe on*. This popular treatise, which ran to thirteen further editions between 1572 and 1656, attempted to enlarge the experience of English gardeners by revealing what was being done abroad, in France and in the Netherlands, where the science of horticulture was receiving greater attention.[4]

Despite the claims made for the Sussex Mascall's pippins, Kent, not Sussex, probably grew the first English ones. Henry VIII's fruiterer Richard Harris established a fruit collection at Teynham, Kent, importing from France many "graftes" of pippins, "before which there had been no pippins in England". His orchard, with royal backing, was described in a 1609 account as "the chief mother of all other orchards for those kinds of fruit in Kent and divers other places".[5] In stimulating fruit-growing around Teynham, Harris had helped to turn this part of Kent into "the Garden of England." Kent's apples and cherries thrived especially around Maidstone and in north Kent between Rainham and the woodland of the Blean. The arrival of many Low Country refugees to eastern parts of the county in the 1560s and 1570s stimulated the London trade in fruit as well as vegetables such as cabbages, carrots and celery. Before long, many refugee families moved nearer their main markets, setting up in Wandsworth, Battersea and Bermondsey.[6]

But in Sussex, further from London markets, apples continued to flourish mostly in small orchard plots attached to large country houses, vicarages and cottages – or in town gardens. At Winchelsea, for example, surviving records show that many dwellings or "messuages" had gardens and orchards; one document, for instance, refers to "Daniel White gent's garden or orchard called Paradise Garden". Orchard expansion was also under way: "Empty plots, stone buildings and walls, formerly a messuage called Harrowes Place, lately burnt down and demolished, with four joining gardens now converted into an orchard in Quarter 20."[7]

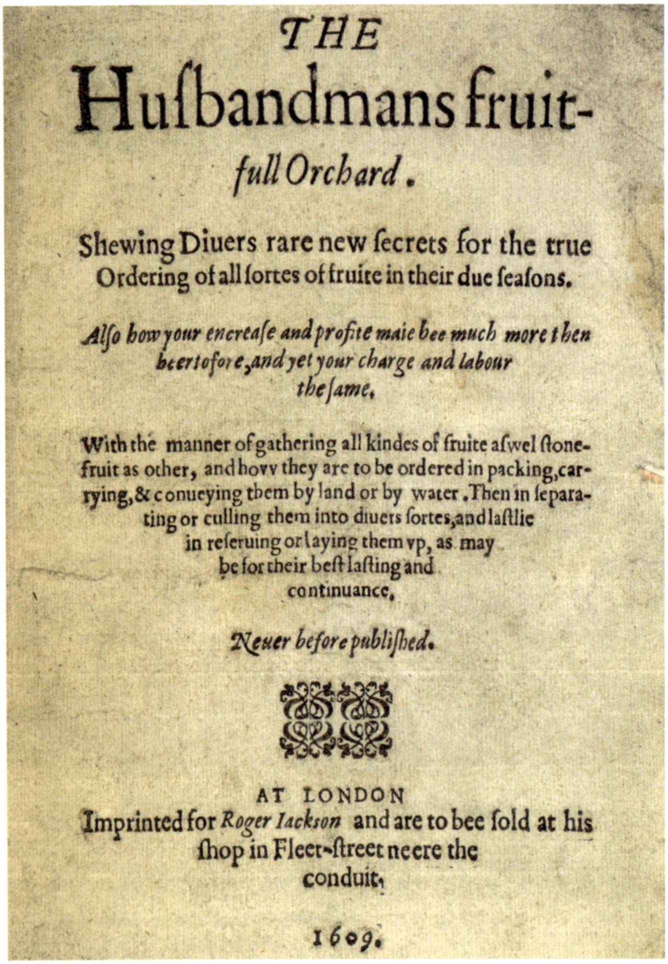
THE
Huſbandmans fruit-
full Orchard.
Shewing Diuers rare new ſecrets for the true Ordering of all ſortes of fruite in their due ſeaſons.
Alſo how your encreaſe and profite maie bee much more then heertofore, and yet your charge and labour the ſame.
With the manner of gathering all kindes of fruite aſwel ſtone-fruit as other, and hovv they are to be ordered in packing, carrying, & conueying them by land or by water. Then in ſeparating or culling them into diuers ſortes, and laſtlie in reſeruing or laying them vp, as may be for their beſt laſting and continuance.
Neuer before publiſhed.
AT LONDON
Imprinted for *Roger Iackson* and are to bee ſold at his ſhop in Fleet-ſtreet neere the conduit.
1609.

Domestic wealth: The Husbandman's fruitful orchard (1603)

By Charles I's reign, as the pastoral vision of landscape became fashionable, wealthy landowners took greater interest in English apples as well as imported varieties. The first botanic garden in England, established at Oxford in 1621, stimulated the cataloguing and exchange of varieties. Queen Henrietta Maria created a lavish garden at Wimbledon Manor well-stocked with a diversity of apples. The Stuart court presented elaborate fruit banquets, often taken in the separate banqueting house or in the garden arbour, where fruit, sweetmeats, jams, pippin jellies, tarts and cream, sweetened with sugar imported from the colonial plantations, were eaten amid pleasing vistas of orchard and gardens.

But apples were for commoners too, and many labourers' cottages had gardens with one or two apple trees alongside vegetable plots.[8] In Sussex, apple orchards regularly featured in documents known as glebe terriers, detailing the land allocated for the upkeep of the clergy within each parish. Orchards were ubiquitous in Dallington and Hastings Deaneries, in far eastern stretches of the county, and in western Sussex too: documents dating from 1635, for example, listed an orchard, hemplott and herb garden at Aldingbourne; a herb garden and orchard at

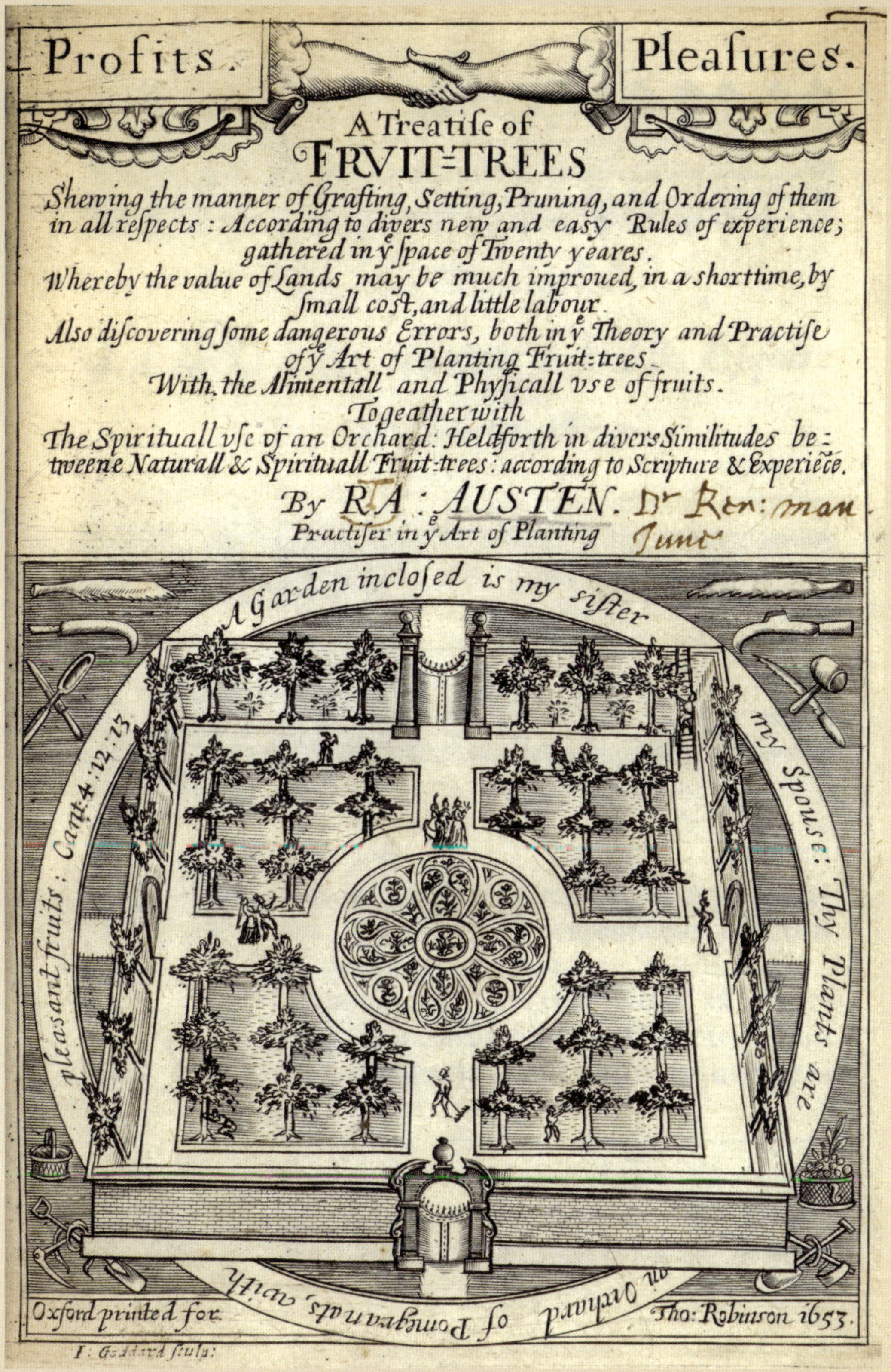
Profits. Pleasures.

A Treatise of
FRVIT-TREES

Shewing the manner of Grafting, Setting, Pruning, and Ordering of them in all respects: According to divers new and easy Rules of experience; gathered in yͤ space of Twenty yeares.

Whereby the value of Lands may be much improued, in a shorttime, by small cost, and little labour.

Also discovering some dangerous Errors, both in yͤ Theory and Practise of yͤ Art of Planting Fruit-trees.

With the Alimentall and Physicall vse of fruits.

Togeather with

The Spirituall vse of an Orchard: Held forth in divers Similitudes betweene Naturall & Spirituall Fruit-trees: according to Scripture & Experiēce.

By RA: AUSTEN. Dr Ren: man
Practiser in yͤ Art of Planting June

A Garden inclosed is my sister my Spouse: Thy Plants are an Orchard of Pomegranats with pleasant fruits: Cant. 4:12:13

Oxford printed for Tho: Robinson 1653.

I: Goddard sculp:

Radical puritan with a reforming eye on orchards: Ralph Austen, A Treatise of Fruit Trees (1653)

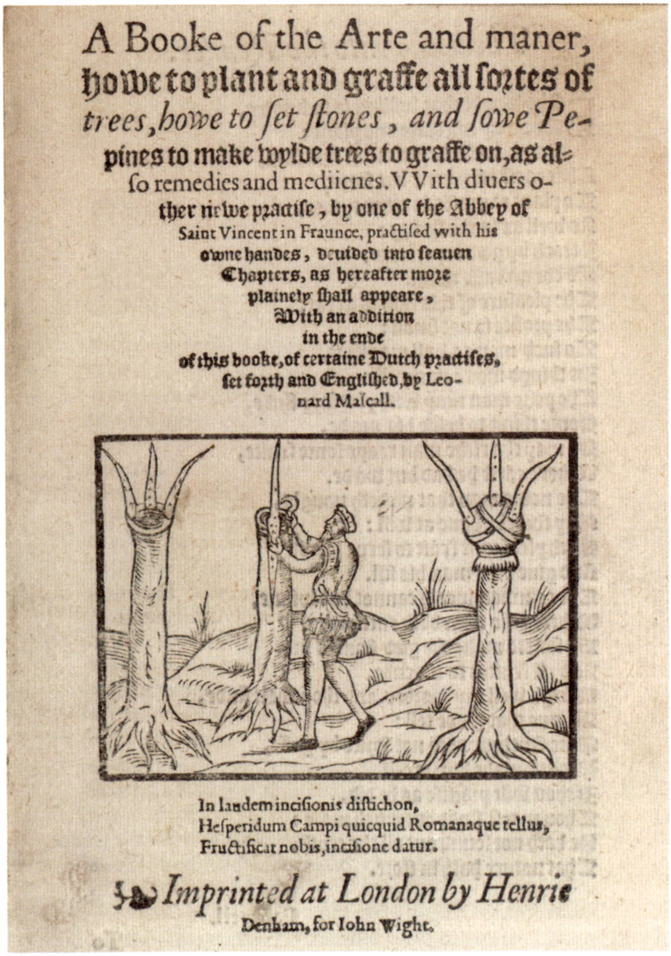
A Booke of the Arte and maner, howe to plant and graffe all sortes of trees, howe to set stones, and sowe Pepines to make wylde trees to graffe on, as also remedies and mediicnes. VVith diuers other newe practise, by one of the Abbey of Saint Vincent in Fraunce, practised with his owne handes, deuided into seauen Chapters, as hereafter more plainely shall appeare, With an addition in the ende of this booke, of certaine Dutch practises, set forth and Englished, by Leonard Mascall.

In laudem incisionis distichon,
Hesperidum Campi quicquid Romanaque tellus,
Fructificat nobis, incisione datur.

Imprinted at London by Henrie Denham, for Iohn Wight.

A Booke of the arte and maner, howe to plant and graffe all sortes of trees by Leonard Mascall, 1569 - the burgeoning book-buying public sought advice on planting orchards and looking after the trees

Ashington; and apples being received as tithe payments (together with pears, hemp and hops) at Bepton and Cocking.[9] An orchard and garden of three-quarters of an acre next to the dwelling house at Slindon was noted in 1635, and in 1663 an orchard among glebe lands and buildings at Storrington.[10]

As the radical Protestant movement in England grew in influence, the cultivation of orchards offered spiritual as well as practical benefits. In 1653 the influential Calvinist and Oxford horticulturist Ralph Austen (c.1612-1676) published *Treatise of Fruit Trees* with a devotional tract, *The Spiritual Use of an Orchard*. His appeal to Oliver Cromwell for a state-sponsored programme of fruit-tree plantings was unsuccessful, but he influenced ideas on horticulture, new crops and methods of production well beyond London, not least through his correspondence with Samuel Hartlib (c.1600-1662).

Austen was known especially for his cider factory, "believing that cider would promote health, allay sedition, and eliminate the call for imported wine," and his customers included Isaac Newton.[11] Cider was generally approved of by Puritans; the apple was seen as a simple, wholesome fruit, with no particular Catholic or "Mediterranean" associations. Cider was greatly preferred to the imported European wines, and would, according to the poet John Philips in his 1708 poem *Cyder*, "please all tasts, and triumph o'er the Vine".[12]

Orchards for ordinary people were also advocated by the Herefordshire parson and agricultural reformer John Beale, who listed the better cider varieties in *Herefordshire orchards* (1657). He also corresponded with the Surrey landowner John Evelyn who shared his interest in orchards; Evelyn's treatise on apples *Pomona* (with a contribution by Beale) was added as an appendix to his well-known book on forestry, *Sylva* (1664).

Together, Beale, Hartlib, Austen and Evelyn constituted the seventeenth-century "fruit lobby" – or as they referred to themselves, those with "hortulan interests". The group's literary outputs were allied with the more practical skills of the newly-arrived Flemish refugees in southeast England.[13] Another enthusiast within this circle was the

Petersfield-born John Worlidge (d. 1693). His *Systema agriculturae, or, The Mystery of Husbandry Discovered* (1669) included a section on cider, later expanded and published as *Vinetum Britannicum, or, A Treatise of Cider* (1676):[14]

> There is no fruit tree in this whole island of Great Britain that is so universal as the apple tree: there being but few places and but little and wherein it delighteth not; hardly any place so cold or so moist, hot or dry, but will thrive and bear fruit.[15]

Worlidge went on to publish *The most easie method for making the best cider* in 1687.

As the status of orchards rose, in the wake of support from influential men, including members of the newly founded Royal Society, more fruit trees were planted across the country, on smallholdings and large estates alike. The orchard boom included Sussex but was most pronounced in Worcestershire, Hereford, Gloucestershire and Shropshire – where new orchards were being made out of former pasture, and cider took on a particular commercial significance.

During the seventeenth century, orchards expanded on farms of every size: the fruit wood provided useful timber, and the apples could be stored during the long winter months. The fashion for planting fruit trees for home and market spread from the gentry to the yeomen farmers, whose orchards were often planted on "mean land", in odd corners and in gardens near to houses, and among whom cider drinking remained highly popular.[16] Smaller orchards produced local varieties, suited to particular soil and climate conditions, and provided cider, cooking and eating apples as well as other fruit. One valued variety was the Cat's Head apple, a somewhat angular and odd-shaped apple with a sharp taste, enjoyed in Kent and Sussex by farmers who took it, cooked and wrapped in dough, with them to their fields.

Across the west Sussex Wealden clay plain, apple-growing was patchy; one historian estimates that 10 per cent of the population of Kirdford had cider presses.[17] In the eastern Weald, orchards were probably more common: surviving documents mentioned orchards at Salehurst in 1588, Etchingham in 1597 and Hailsham in 1625. An estate in Bexhill, Langney and Westham in 1606 possessed as many as 40 separate orchards in 1620.

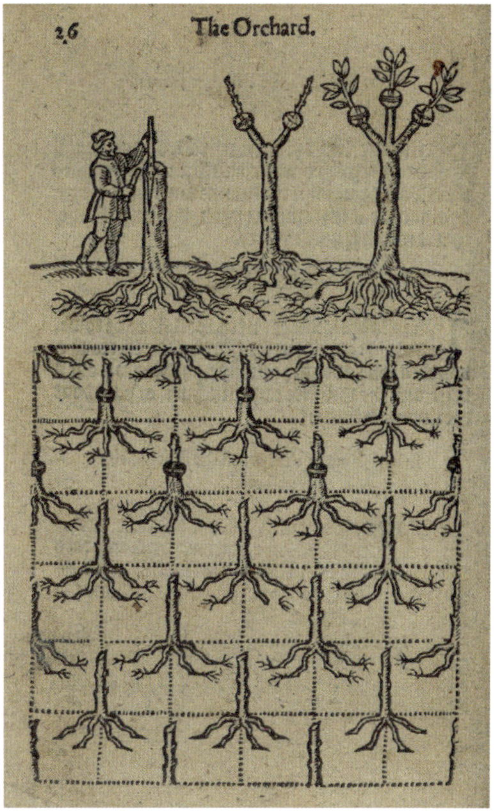

The Orchard and the Garden: containing certaine necessarie, secret, and ordinarie knowledges in Grafting and Gardening. (London: Printed by Adam Islip, 1602) – illustrations depict how to graft, the layout of young trees in an orchard, and the specialized tools used

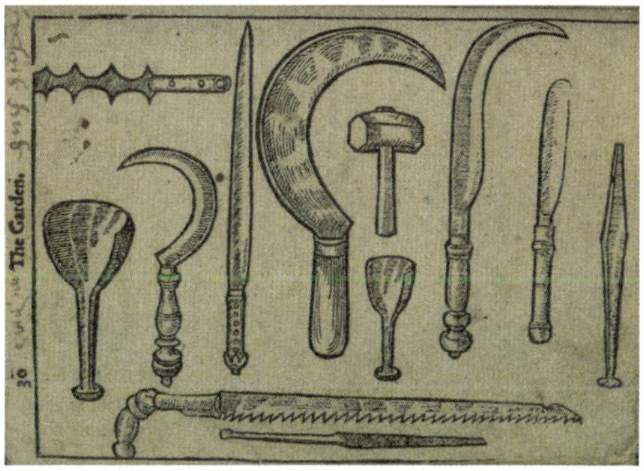

In eastern Sussex, growers could benefit by proximity to the Wealden orchards in neighbouring Kent – in the 1660s, trade in apples down the Medway valley to London was flourishing. In Ashdown Forest there was an orchard at Old Lodge in 1658. At nearby West Hoathly Thomas Comber was reported to the Lord Chancellor Ashley in May 1671 for withholding tithes on his produce, which included great quantities of honey and wax as well as "great store of apples, peares, plumes, walnuts and other fruites". A witness in the case stated that Comber had sold two bushels of walnuts in 1670, worth 12s-14s.[18]

Most orchards were modest in size and often situated near the farmhouse. For example, the Rogers family at Iford, near Lewes, in January 1658 rented from Swanborough Manor a cottage, barn, garden, orchard and land of 10 acres.[19] While much Wealden produce was consumed as fresh fruit, much was still being converted to cider (and perry from pears). One document from 1609 cited "apple mylls", used to crush apples for cider, stored in an outbuilding at Robertsbridge manor. In 1657 the Rev. Giles Moore of Horsted Keynes noted in his journal his purchase from London of a selection of trees: apricot, an "orange pare tree", a "pare royal or Windsor pare", two Kentish pippins and two "Flanders cherryes". Later he paid a neighbour 3/6d to purchase the hay growing in his orchard. In 1673 Moore "payed Thos. Ward, beysides his diet, for four dayes worke about the garden and orchard, 4s."[20]

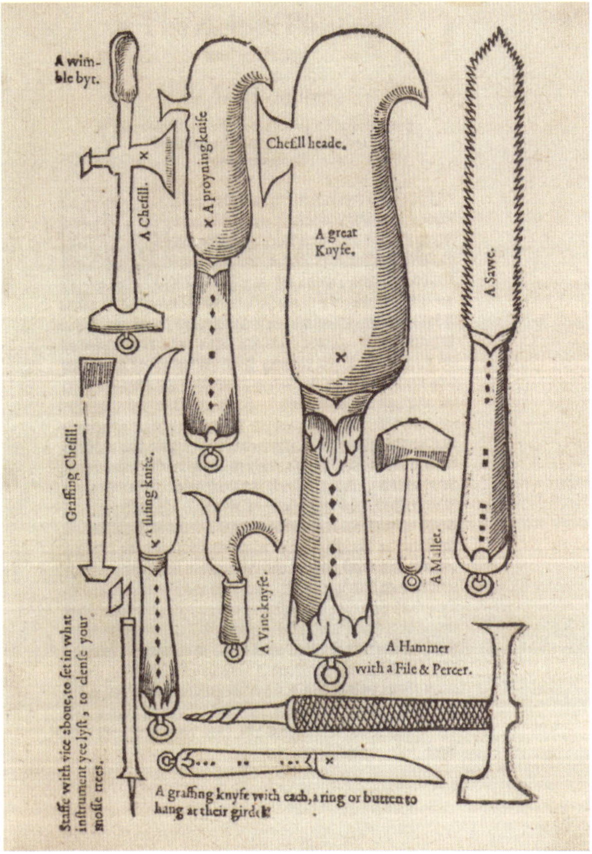

Best-seller: Leonard Mascall's handbook, published 1572, with its detailed advice on planting and grafting an orchard, was a great publishing success, running to fourteen editions over five decades. The extensive range of tools and equipment on show testifies to the fruit-growing skills of his age

Orchards and cider-making equipment were valuable, and cited time and again in probate inventories – listings of the property and possessions of the deceased for probate purposes – dating from the seventeenth century and on into the eighteenth. Documents from the 1600s make repeated reference to apple lofts, where the apple harvest was stored until ready for cider-making, for instance at Rudgewick and Billingshurst. Apples were being grown and sent from West Hoathly in 1672.[21] At Amberley, Richard Hopkins' probate inventory, dated January 1693, left "1 cyder press and trough etc. worth £1/10/0". Cider makers of the day broke up the apples in wooden troughs, then crushed them in a press to extract as much juice as possible, which was then strained and left to ferment naturally in large barrels or hogsheads. Agnes Pearley, widow from Bignor, in March 1690 left "1 cyder press and 2 cheese presses £2". Richard Mantell, a gardener at Southover, Lewes, left "4 glass frames and a parcel of apples 10s", and a "cyder room" with cider press in October 1720. John Lullham from Sedlescombe, described as a yeoman in August 1718, left "apples upon the trees £1". Unfortunately, no acreages are given for orchards in the inventories.[22] In the same year the inventory for Samuel Padie, rector of Wiston 1670-1718, recorded: "In the barn ... a cider mill and press and frame to hold apples and a small ladder £2/10s." John Dodderidge, vicar of Whatlington 1685-1734, left "one apple trofe".[23]

One Sussex man whose knowledge of apples led him to perfect a way of making cider into a drink to rival sherry and port was Richard Haines of Sullington (1633-1685).[24] In 1684 he published his manifesto for the new beverage: *Aphorisms upon the new way of improving cyder, or making cyder-royal, lately discovered ... To which are added, certain expedients concerning raising and planting of apple-trees, gooseberry-trees, &c. with respect to cheapness, expedition, certain growing, and fruitfulness, etc.*

Haines' treatise set forth his techniques for making "cyder-royal" according to a patent of 1684, granting him sole license to his technique for 14 years. Ordinary cider at this time was typically only moderately alcoholic, diluted with water during its production and with a strength on a par with a beer or ale. A quart of such cider might typically have been drunk in a day. Yet Haines believed that, properly made, his cyder-royal could stand comparison with the best French wines. His book described how to make a fortified wine, like port or sherry, by distilling cider to make a sort of cider brandy, and then adding that strong spirit to cider. He emphasised the good keeping qualities of cyder-royal, and indeed, brandy was first added to local Portuguese wines to make port, to help to stabilise the wines being shipped to England.

His book gave advice on the economics of production, advocating the planting of Redstreak, a variety now lost but the most popular cider apple in Haines' day, as well as the Golden Pippin, a venerable Sussex variety still available today. In southern counties such as Sussex, Golden Pippin was particularly esteemed for producing a delicious liquor ideal for cider-making, which was more often made from whatever mixture of surplus eating and cooking apples happened to be available.

Haines gave good and detailed advice on setting out an orchard:

> What and how to plant: One acre of land planted with apple trees may bring in 8d worth of cider-royal, though its price be but 1d. a quart. Plant thus: Eight score Red-Strakes and Golden Pippins per acre 16 feet apart, which at only 1 bushel per tree will bear 160 bushels per acre. Each hogshead of cider requires 20 bushels of apples, and 8 hogsheads of ordinary cider will make 4 of cider-royal. This at Id. per quart will make 2 per hogshead.[25]
> By planting the trees 20 feet one way and 12 the other you can plough between and plant grass or corn. But gooseberries and currants so

An apple with many virtues: the Golden Pippin, in watercolour by Nicky Ashworth

> planted will be better, and will yield 4 hogsheads of wine-royal, which again at 2d. per quart will give 8 more per acre.
>
> PROFIT ON FRUIT-GROWING. But apple trees usually bear from 4 to 20 bushels a tree, and so, taking the lower number, we may get 16 hogsheads instead of 4.... Again, gooseberries and currant trees, as experience teaches, well husbanded one with another, may yield a gallon apiece. As 16 trees will stand on one rod of land, 4 feet apart, one rod will thus produce 2 bushels of fruit. There being 160 rods in an acre, the total yield will be 320 bushels, i.e., 8 hogsheads of wine.

All this would, Haines claimed, make land that was of marginal value previously come to some account. Furthermore:

> Being the product of our own soil, it must needs suit our constitutions better than outlandish liquors. It can be recommended as an appetizer and tonic, and even excessive indulgence has no harmful effects. A home-brewed article, consumed at home, it will save us half a million pounds yearly.

As to the apple varieties:

> BEST FRUIT FOR THE PURPOSE. The Red-Strake, a kind of wilding, is suitable, but the best apple for the table and for cider too is the Golden Pippin, a quick grower and good bearer, yielding the most liquor and the best. The very husks of the latter, even when cider has been made from the fruit, will yield more spirit than others. Nor is it any more difficult to raise good fruit than bad.
>
> It is certain that good wildings and good crabs are better for cider than the most delicious summer or winter table fruit or sweet apples (the Golden Pippin alone excepted). The bitter sharp crab is much better than a bitter sweet apple, because its juice will give twice as much spirit. The sweet and sour tastes are left behind in the earthy phlegmatic part of the cider. The spirit of sour cider will not be sweetened by any amount of sugar.

Modern cider makers still distinguish between bittersharps and bittersweets – reflections of the mix of acid and sugar and tannin in varieties of cider apple – and take care to ensure the right mix of apples goes into their cider to achieve the taste they desire.

Unfortunately, Haines' idea of making a good strong up-market cider on a par with wine or cider-brandy never really took off, and as a result England today has no tradition of home-grown calvados to rival France. Home-grown fortified cider-wine and cider-brandy evidently couldn't compete with foreign imports – especially French wine and brandy, rum from the West Indies, and cheap gin from Holland.

Nonetheless, apple-growing and cider-making continued to thrive in Sussex, though not all orchards were harmonious scenes of serene planting and harvesting. In 1610 Thomas Brewer had published *The bloudy mother, or The most inhumane murthers, committed by Iane Hattersley vpon diuers infants, the issue of her owne bodie*. One infanticide was said to have taken place at East Grinstead in July 1609, with the title page of Brewer's tract depicting the burial of three children in the orchard. Described by Brewer as "more than tygerlike" and without feelings, "a heart of steele and eies of marble," Jane Hattersley believed that her lover Adamson would marry her when his own wife died. When Adamson sold the orchard, neighbours became suspicious and unearthed many small bones. Jane was executed, but Adamson was acquitted only to die later a lonely death in bed, "devoured by wormes and lice". [26]

The period from 1500 to 1700 saw a remarkable increase in expertise and experimentation in apple-growing, as interest in apple varieties burgeoned, and more and more people ate fresh fruit as well as consuming the harvest as cider. The east of

The Bloudy Mother,
OR
The moſt inhumane murthers, committed
by Iane Hatterſley vpon diuers Infants, the iſſue of her
owne bodie: & the priuate burying of them in an Orchard
with her Araignment and execution.
As alſo, The moſt loathſome and lamentable end of Adam
Adamſon her Maſter, the vnlawfull begetter of thoſe vnfortunate
Babes, being eaten and conſumed aliue with Wormes and Lice.
At Eaſt Grinſted in Suſſex neere London, in Iuly laſt. 1609.
Printed for Iohn Buſbie, and are to be ſold by Arthur Iohnſon in
Paules Churchyard at the ſigne of the White Horſe.

Sussex in particular became an established centre of orchard produce, and cider-making thrived throughout Sussex, ensuring that orchards were common features in the county's landscape.

Orchard crime scene: in 1610, Thomas Brewer published a gruesome tale of multiple infanticide, after three babies' bodies were found buried in an orchard in East Grinstead

Endnotes for Chapter 4

1 Joan Thirsk, *Alternative agriculture: a history* (OUP 1997), 18-31.

2 W. Lambarde, *A perambulation of Kent* (1576), 8.

3 M.A. Lower, *The Worthies of Sussex* Lewes: Geo P. Bacon 1865, 52-4.

4 John Considine, 'Mascall, Leonard (d. 1589)', *Oxford Dictionary of National Biography*, Oxford University Press, 2004 [http://www.oxforddnb.com/view/article/18256, accessed 9 March 2011]. The writer states that this Leonard Mascall was not the Mascall of Plumstead [a mistaken version of Plumpton] in Sussex, since the latter was from a family to which the writer was not related. So the Mascall credited with introducing carp and pippins into England was not the Mascall who wrote *the Arte of and Manner howe to plant and graffe all sortes of trees*. The Mascall who is the subject of the

DNB entry was buried at Farnham Royal, Buckinghamshire, in early May 1589.

5 Joan Thirsk, 'Farming techniques' in J. Thirsk (ed), *The Agrarian History of England and Wales IV 1500-1640* Cambridge: Cambridge University Press 1967, 196; T. Hill, The profitable art of gardening 1574., cited in H.Taylor, *The apples of England* London: Crosby Lockwood and Son 1948 3rd edtn, 11.

6 R. Webber, *The early horticulturalists* (1968), 19.

7 ESRO WIN/56, Court Book of the Court of Record 1604-1637. The quarters referred to are the blocks of building land within Winchelsea's medieval town walls.

8 A. Everitt, 'Farm labourers' in J. Thirsk (ed), , *The Agrarian History of England and Wales Vol. 4, 1500-1640* (1967), 452.

9 British Library Add Mss 39,467. Terriers of Livings. Chichester Registry. Dunkin Collection vol. 142.

10 West Sussex Record Office PAR 209/1/1/2; EP 1/25/3; EP 1/25/3.

11 The full title was A *treatise of fruit-trees, shewing the manner of grafting, planting, pruning, and ordering of them in all respects ... discovering some errors in the theory, and practise of this art to be avoyded : with the alimentall, and physicall use of fruits : to which may be annexed the second part, viz. The spiritual use of an orchard ...*, Oxford : Printed by Henry Hall ... for Thomas Robinson (1657). J. G. Turner, 'Austen, Ralph (c.1612–1676)', *Oxford Dictionary of National Biography*, online edn, Oxford University Press, Sept 2004 [http://www.oxforddnb.com/view/article/905 [accessed 16 July 2010].

12 Quoted in J. Russell, *Man-made Eden: historic orchards in Somerset and Gloucestershire* Bristol: Redcliffe Press 2007, 44.

13 See also D. Chambers, '"Wild pastorall encounter"; John Evelyn, John Beale and the renegotiation of pastoral in the mid-seventeenth century' in M. Leslie and T. Raylor (Eds), *Culture and cultivation in early modern England: writing and the land* Leicester: Leicester University Press 1992, 173-94.

14 Ernest Clarke, 'Worlidge, John (d. 1693)', rev. Anita McConnell, *Oxford Dictionary of National Biography*, Oxford University Press, 2004 [http://www.oxforddnb.com/view/article/29975, accessed 16 July 2010]

15 J. Worlidge, *Vinetum Britannicum: or, a Treatise of cider ... Together with the method of propagating all sorts of vinous fruit-trees* 1676 [published under the name J.W. Gent.]

16 Thirsk, *Alternative agriculture*, 39-58.

17 G.H.Kenyon, 'Kirdford inventories, 1611-1776', *Sussex Archaeological Collections*, 93 (1955), 117.

18 East Sussex Record Office AMS 3305-7: complaint by John Hull, clerk [vicar] of West Hoathly, Sussex.

19 East Sussex Record Office, Adams Mss 46.

20 R. Bird, *The Journal of Giles Moore 1656-1679* Lewes: Sussex Record Society 1971,115, 263.

21 B. Short, 'The South East', in J. Thirsk (ed), *The Agrarian History of England and Wales Vol. 5,1640-1750*, 273.

22 WSRO Amberley/69; Bignor/19; ESRO Inv 1256; 1718/989.

23 A Hughes, *Sussex clergy inventories 1600-1750* SRS 91, 2009.

24 C.R. Haines, *A complete memoir of Richard Haines (1633-16850, a forgotten Sussex worthy, with a full account of his ancestry and posterity* (1899).

25 A hogshead of ale or beer is 54 gallons, while a Hogshead of wine is 63 (old Winchester wine) gallons, each of which was 0.8331 Imperial gallons, making a hogshead of wine 52.5 Imperial gallons.

26 S. Clark, *Women and crime in the street literature of early modern England*, London: Palgrave 2003, 40, 49-52.

5. 1700–1840

5

Apples in Sussex: 1700-1840

"Good apple pies are a considerable part of our domestic happiness."
Jane Austen.[1]

In the age of great estates, apple orchards consisted of tall standard trees, with grass below grazed by sheep or cattle. Apple puddings and pies in particular were much favoured, sweetened with sugar imported from the Caribbean. For farming families it was boiled apple dumplings.[2]

Apple varieties favoured at this time were the small multi-purpose Golden Pippin, a variety with Sussex links, and the sharply flavoured and long-keeping Nonpareil. Both might feature at the end of meals, in "the dessert". Also popular were Golden Reinettes, as well as French varieties imported during a century in which the wealthy travelled widely on the Continent.

While neighbouring Kent established its reputation as the market garden of London, supplying orchard fruit for the table, the West Country concentrated on making cider. Sussex did quite a bit of both, growing fresh fruit for market and supporting many a cider-maker across the county. Though near to London, Sussex before the railways was still relatively cut off from the capital, and much of Sussex's orchard produce was made into cider for local consumption, with fresh apples serving local needs.

By and large, the orchards in Sussex were small, designed to supply the mansion, vicarage or farmhouse. The vicarage of Hartfield, for example, in 1724 had two acres of glebe land attached, "which is most of it in an orchard".[3] Edward Stuart, rector of Wiggonholt, between Parham and Pulborough, recorded inside the back cover of the Wiggonholt parish register that in 1713: "I planted a new orchard at Greatham between the parsonage and the barn," and over the next two years "planted ye field adjoining to my garden with apple trees."

On country estates, walled gardens became increasingly popular. Laid out with a tidy array of paths and avenues, maintained by an army of gardeners, these enclosures featured fruit trees trained into fans and espaliers against walls or fences. The apples flourished in the protected environment, ready for the dining table or banquet. Thus arose the great fruit and vegetable gardens of this period. For instance, at West Dean, near Chichester, a grand walled garden with fruit trees was laid out in 1810. At East Hoathly a maze was planted in 1821 which contained a variety of fruit. Apple, pear, plum, cherry and greengage trees were planted at Prospect House (also known as Hesmonds) for Matthew Martin. A plan drawn up at this time shows the outline of the maze and lists the different species of fruit.[4]

In the great houses apples could be consumed on a grand scale: the Duke of Richmond's estate at Goodwood paid £1.8s at the end of November 1746 for 10 sacks of apples. Cider remained an important commodity as well. In March 1747 freight and wharfage of 19 shillings was paid on two hogsheads of West Country cider which had been sent via Exeter and Emsworth.[5] By contrast, the Duke of Somerset's estate at Petworth was selling apples in 1739: a "load of apples [was] sold to Mr Hill of Chichester" for £4.0.6d.[6]

In eastern Sussex, the Bishop of Durham, Richard Trevor (1707-1771) took possession of Glynde Place in 1743. He lavished attention on the house and surrounding land. In setting out his garden at Glynde Place in 1754, he produced a detailed inventory of the fruit trees sent for planting from Henry Hewitt and Co., Brompton, where plant nurseries flourished well into the eighteenth century. Besides such things as dwarf peaches, greengages, plums, Morello cherries, pears and grapes, the inventory of purchases listed 48 apple trees. There were four trees each of Codlings, Golden Pippins and Nonpareils, and two each of a large selection of varieties, many of which are now lost: Holland Pippins, Royal Russets, Joan's Pearmains, Royal Pearmains, Margills, Aromatic Pippins, Kirton Pippins, Chester Pearmains, Golden Rennets, Barnard's baking apples, Pie pie, Red Colville, Margretings, Monstrous Rennets, Quince apples, Spencers Pippins and Nonsuch.[7]

Apples also flourished just down the road from Glynde, at the village of Ringmer, where the naturalist and Hampshire vicar Gilbert White (1720-1793) was a regular visitor to his aunt, Mrs Rebecca Snookes, who lived just half a mile from the South Downs. In his *Naturalist's Journal*, White mentioned the crossbills visiting her "Scotch Pines" on 5 October 1770 and later noted, on 9 September 1777: "fern-owls [nightjars] haunt Mrs Snooke's orchard in autumn." Despite being on the clay-rich low weald, apples grew well in Ringmer, he concluded. "Ringmer soil not clay at top but brick-loam: bears good apples, pears, & grapes. Clay under, which holds water like a dish. The trees are mostly elms," he wrote on 8 November, 1771. And in his entry for Sept. 26-Oct. 7, 1778, he commented: "Mrs. Snookes has gathered-in all her apples, & pears: her fruit is finely flavoured in such hot years. Mrs Snooke's black grapes begin to ripen. No wasps here."

White famously took a particular interest in his aunt's ageing tortoise, Timothy, who lived in her walled courtyard with a well-stocked herbaceous border. Mrs Snookes lovingly looked after Timothy over 40 years, feeding him home-grown cucumbers and

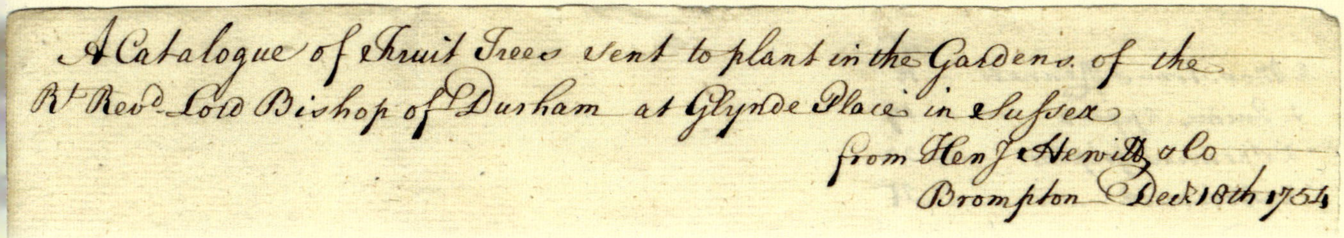

A Catalogue of Fruit Trees sent to plant in the Gardens of the
Rt Revd Lord Bishop of Durham at Glynde Place in Sussex
from Henry Hewitt & Co
Brompton Decr 18th 1754

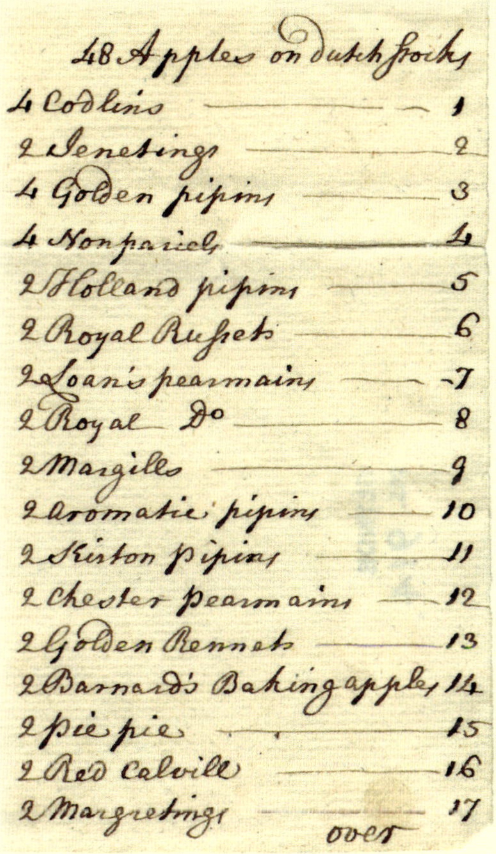

48 Apples on dutch stocks

4	Codlins	1
2	Jenetings	2
4	Golden pipins	3
4	Nonpareils	4
2	Holland pipins	5
2	Royal Russets	6
2	Loan's pearmains	7
2	Royal Do	8
2	Margills	9
2	Aromatic pipins	10
2	Kirton pipins	11
2	Chester pearmains	12
2	Golden Rennets	13
2	Barnard's Baking apples	14
2	pie pie	15
2	Red Calvill	16
2	Margretings	17

over

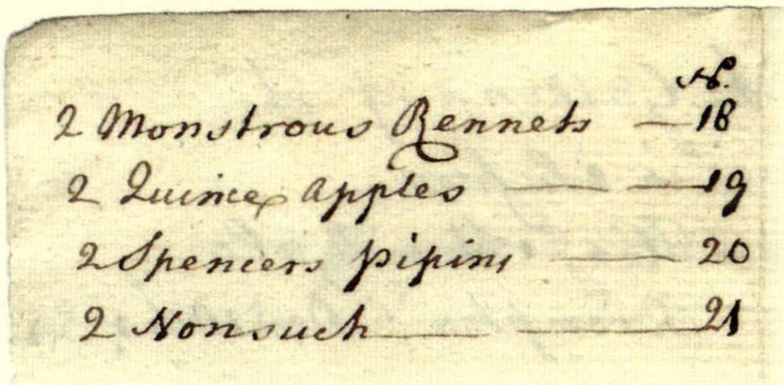

		No.
2	Monstrous Rennets	18
2	Quince apples	19
2	Spencers pipins	20
2	Nonsuch	21

The 1754 inventory of fruit trees sent to Glynde Place from Henry Hewitt & Co. of Brompton. Extracted here from a very lengthy listing are the entries detailing forty-eight apple trees

kidney beans, and when she died, aged 86, in 1780, White took Timothy to live with him in Selborne, where he lived on another 14 years – outliving White by a year. At Selborne, Timothy enjoyed a rich diet, putting on weight and spending much time in White's vegetable garden and a warm spot in "the border under the fruit-wall," where White grew gooseberries, peaches, pears and apples.

Books on orchards continued to appear in print throughout the eighteenth century, sometimes specifically focused on methods of cultivating fruit. In 1732 an anonymous publication in London, for example, was entitled *Essay concerning the best method of pruning fruit trees; also the method of pruning timber trees; and also a discourse concerning the improvement of the potatoe*. Or there was the intriguingly titled *Adam's luxury and Eve's cookery; or the kitchen garden displayed* (London 1744).[8] In 1754 William Ellis (c.1700-1758) of little Gaddesden, Hertfordshire, published *The complete cyderman*, with advice drawn from west-country cider-makers in Devon, Herefordshire and Cornwall.

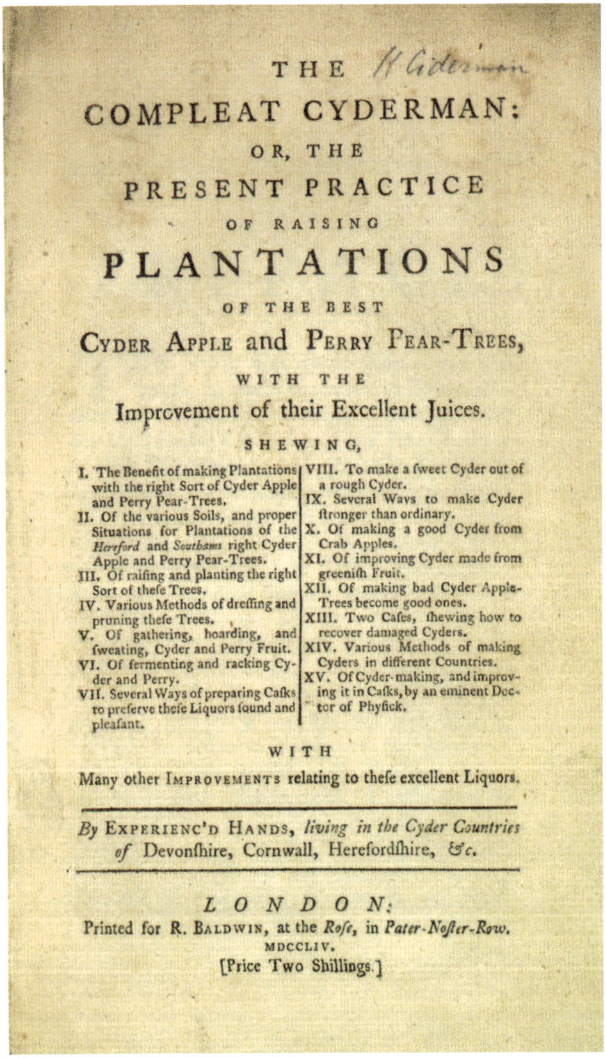

THE

COMPLEAT CYDERMAN:

OR, THE

PRESENT PRACTICE

OF RAISING

PLANTATIONS

OF THE BEST

CYDER APPLE and PERRY PEAR-TREES,

WITH THE

Improvement of their Excellent Juices.

SHEWING,

I. The Benefit of making Plantations with the right Sort of Cyder Apple and Perry Pear-Trees.
II. Of the various Soils, and proper Situations for Plantations of the *Hereford* and *Southams* right Cyder Apple and Perry Pear-Trees.
III. Of raiſing and planting the right Sort of theſe Trees.
IV. Various Methods of dreſſing and pruning theſe Trees.
V. Of gathering, hoarding, and ſweating, Cyder and Perry Fruit.
VI. Of fermenting and racking Cyder and Perry.
VII. Several Ways of preparing Caſks to preſerve theſe Liquors ſound and pleaſant.
VIII. To make a ſweet Cyder out of a rough Cyder.
IX. Several Ways to make Cyder ſtronger than ordinary.
X. Of making a good Cyder from Crab Apples.
XI. Of improving Cyder made from greeniſh Fruit.
XII. Of making bad Cyder Apple-Trees become good ones.
XIII. Two Caſes, ſhewing how to recover damaged Cyders.
XIV. Various Methods of making Cyders in different Countries.
XV. Of Cyder-making, and improving it in Caſks, by an eminent Doctor of Phyſick.

WITH

Many other IMPROVEMENTS relating to theſe excellent Liquors.

By EXPERIENC'D HANDS, *living in the Cyder Countries of* Devonſhire, Cornwall, Herefordſhire, *&c.*

LONDON:

Printed for R. BALDWIN, at the *Roſe*, in *Pater-Noſter-Row.*

MDCCLIV.

[Price Two Shillings.]

William Ellis gave extensive advice to would-be cider and perry makers in his treatise, The Compleat Cyderman, published in 1754

In the second half of the eighteenth century, as interest grew in systematising the identification of apple varieties, many professional gardeners published their own thoughts on the subject. The "aspirant clerical botanists" in Georgian England also began to take more of an interest in the "rustic" knowledge of the countryside. More formal investigations were stimulated by the founding of the London Horticultural Society in 1804, the forerunner of today's Royal Horticultural Society (RHS).

Pomologists struggled to document the hundreds of apple varieties that existed but about which they knew very little. Even deciding quite how many genuine varieties there were wasn't easy, given that the same apple might be known by a variety of different names around the country. What's more, many local apple varieties were extremely local, having emerged from wayside trees, bird droppings or discarded fruit, and grown only within a small region.

The systematisers sought to assess the quality and range of apples, and to identify and select the most promising varieties, such as Ribston Pippin, Golden Noble or Blenheim Orange. As the work of the society lessened confusion over varieties, enthusiasts began to exchange apple-tree cuttings ready for grafting across England and with like-minded enthusiasts abroad. Any variety could be readily propagated by grafting the cutting onto a rootstock, using straightforward techniques that had remained essentially unchanged over many centuries.

As the century progressed, there came a spate of beautifully produced books with illustrations of apple varieties, known as *Pomonas*, after the Roman goddess. George Brookshaw (1751-1823) produced his *Pomona Britannica* (1812) dedicated to the Brighton-loving

George, Prince Regent and future King George IV, which contained 60 plates and described 256 varieties. His lead was followed by others, such as William Hooker, with his *Pomona Londiniensis* (1813-18).

As an aside, Sussex has an apple-themed connection with George IV through his secret wife, Mrs Fitzherbert (1756-1837), who was a very popular figure in Brighton. An apple dish of the day was named "Apple, or Fitzherbert pudding" in her honour. The recipe, reproduced by the English Folk Cookery Association in 1936, follows:

> Three-quarter pound of the pulp of sharp apples boiled, 3 oz of butter clarified, 3 oz of sugar, 4 eggs well beaten, a little nutmeg, lemon peel, orange flour water and grated bread. Mix well together and lay a thin puff paste in the bottom of your dish.[9]

Royal connections: George Brookshaw dedicated his magnificent celebration of appledom, Pomona Britannica, published in 1812, to the Brighton-loving Prince Regent

As a more scientific approach to fruit cultivation gained ground, the government encouraged formal investigations into farming methods and the rural economy. The Board of Agriculture, founded in 1793, promoted a series of surveys by leading authorities, referred to as *General Views*, with Arthur Young at the helm, as secretary to the Board. His son, the Reverend Arthur Young, got the job of producing the *General View* for Sussex, and in 1813 recorded substantial orchards in western Sussex, where cider was "held in much estimation". In the west, the Earl of Egremont's estate at Petworth, and at nearby Sutton, Bury and Bignor, was a focal point for agricultural improvement in the county. Its extensive orchards produced cider for Petworth House. Along with venison, cider featured strongly in gifts made by the earl to stewards, neighbouring landowners and other locally influential people. The apples were thought to thrive best on lighter, sandy soils rather than on the heavier clays, and the varieties were referred to as "pear-apples, maiden apples, cockle-pippin etc". Produce was marketed at Petworth and Midhurst, but most was consumed by the farmers and labourers' families.[10] William Marshall, writing in 1798, also mentioned apple orchards in the western Weald of Sussex, and commented on their importance for cider.[11]

Three Sussex apple varieties – Egremont Russet, Bossom and Petworth Nonpareil – were bred by the third Earl of Egremont and Mr Slater at Petworth. Today, Egremont Russet is still a well-known apple variety. The archives of the estate include correspondence about types of apple with the rector of Petworth, the Rev. Charles Dunster, as well as apple-related nurserymen's bills between 1782 and 1822. In his *General View of the Agriculture of the County of Sussex*, the Rev. Arthur Young noted that the earl had obtained a cider press from Hereford, and the estate archives include correspondence about cider apple varieties. Most of the farmhouses around Petworth possessed small orchards, often scattered across several plots.[12]

Made to last: A cider press was a valuable asset to own or inherit, worthy of being listed among prized property in an individual's will

All the same, among many labouring families, apples and cider continued to form part of an elaborate barter economy in the western Sussex Weald in which little money changed hands. At Lodsworth in the 1820s, Phillip Rapson, a cottage landlord, sawyer and small farmer, rented a cottage to William Leggatt in return for labour or goods, and also made 72 gallons of cider from Leggatt's own apples. Henry Boxall similarly rented a cottage and gave apples to Rapson. Harvest work was supplemented by a cider allowance, which was itself a remnant of the former practice of paying part of a wage in cider, a practice supposedly abolished by the series of Truck Acts in the nineteenth century, outlawing part-payment in kind, but surviving in rural communities at harvest-time especially.[13] Certainly in years when the apple harvest was exceptionally heavy, farm workers were partly remunerated in cider, even where a commercial apple-growing enterprise was concerned. Workers might carry their cider to the fields in a small wooden firkin or stone jar. In June 1838 Robert Tribe did some paving work for Rapson, who paid him partly with 7 quarts of cider, presumably of his own making. Indeed, the Rev. Arthur Young had previously commented that:

> Lodsworth is noted for the excellent flavour of its cider. The Author has had the pleasure of tasting it at the earl of Egremont's table, of a superior quality.[14]

The account books of John Tyler (1773-1857), who lived at White House Farm, Horney Common on the edge of Ashdown Forest, underline the importance of apples in the cottage economy. He was a shoemaker until about 1823, when it appears that he gave up his business to work as a farmer and carrier, thereafter deriving considerable income from the sale of apples and geese.[15] Such multiple employment was typical of many families in the Weald, and orchards had their part to play in bolstering family incomes.

In the 1820s the radical writer William Cobbett investigated the living conditions of poor labouring families in the Sussex Weald. In his book *Emigrant's Guide* (1829) he included a letter home written by a woman who had emigrated to the Hudson River to her father at Battle. Underlining the significance of cider as a drink for labouring families on both sides of the Atlantic, the letter described her new life in the US:

> We live close by a large river… a very fruitful place for apples, cherries, raspberries, grapes, plums; any one may get them without money. … we can buy a

Critical eye on the state of the nation: William Cobbett exposed the plight of the Sussex poor and advocated fruit-growing as an extra source of food and income for countryfolk

> leg of mutton every day, and green pease, and French-beans; and we have got in thirty-two gallons of cider for fourteen shillings… I wish you was all here to help drink it.[16]

Cobbett, in his *Rural Rides*, conducted on horseback in the southeast of England, and first published in 1830, took note of the orchards in Sussex. In riding from Horsham to Petworth in August 1823 he wrote:

> there are a good many *orchards* along in the Weald, and they have some apples this year; but, in general, not many. The apple trees are planted very *thickly*, and, of course, they are small; but, they appear healthy in general; and, in some places there is a good deal of fruit, even this year.[17]

And writing from Singleton:

> The *apple trees* at Donnington show their gratitude to the hill for its shelter; for I have seldom seen apple trees in England so large, so fine, and, in general, so flourishing. I should like to have, or to see, *an orchard of American apples* under this hill… it affords complete *shelter* from the mischievous winds.[18]

Between Singleton and Funtington he thought the orchards were good in general. But in the Lavant area, despite good soils and "prodigious crops of walnuts… the apples are bad along here. The south west winds have cut them off; and indeed, how should it be otherwise, if these winds happen to prevail in May, or early in June".[19] A similar story was told from near Ashurst and Worth:

> The *orchards* no great things along here. Some apples here and there; but small and stunted. I do not know that I have seen to-day any *one tree* well loaded with fine apples.[20]

The orchard could also be seen as an indication of wealth, at least when planted on a large scale. Such landscapes attracted unwelcome attention during the post-Napoleonic war years, a time of great economic distress. In the 1820s and 1830s overt protest among the poor of Sussex led to so-called plant maiming – deliberate damage to trees and crops. The attacks were invariably directed at orchards and timber plantations that had been recently planted and had markedly altered the farming landscape by taking away agricultural land that had provided employment and income for labourers and their families. After one such attack, two labourers were transported for life at the Kent Assizes in 1827 for destroying 69 apple trees and 16 pear trees at Mereworth, while in two other cases, at Crawley, Hampshire, in 1829, men were transported for seven years for destroying 74 "valuable" fruit trees and for the destruction of fir trees at Marlborough. On 4 December 1825 the orchard of William Brooker at Alfriston in Sussex was attacked, with "upwards of 100 choice apple and pear trees in his orchard" destroyed. In November 1828 quantities of shrubs, flowers and fruit trees were destroyed at nearby Ringmer. And in May 1829 a garden and nursery at Rumboldswhyke near Chichester was attacked and "some choice fruit trees cut up".[21]

By the 1830s there were also signs of a growing concern that has continued to beset apple growers and consumers: alarm over the neglect of English home-produced varieties in favour of imported apples. By 1838 120,000 bushels of apples from North America were being imported to the Port of London alone.[22]

The influx of foreign apples prompted a parliamentary enquiry in 1839 into the fresh fruit trade, despite agreement that English apples were of higher quality. Temporary duties had been imposed on imported apples and pears from the Continent more than century earlier, in 1721-2 and 1736, for instance, in a bid to stimulate English orchard planting and domestic markets.[23]

Despite the pressure from imports, apple growing continued as a commercial enterprise in many parts of Sussex. Mr Stevenson of Framfield, in the Low Weald of eastern Sussex, recommended in 1830 – in *The library of agricultural and horticultural knowledge* – six varieties for a stiff, clayey soil: these were Scarlet Pearmain; Winter Pearmain; June-eating or Jenetin; Shepherd's Seedling; Wadhurst Pippin; and Golden Russet. The Nonpareil was also recommended. It was said that these all thrived well on Mr Stevenson's orchard on a rising hill, well protected from southern, south-westerly and north-easterly winds. The soil was a stiff clay. Pippin, Broad-eyed Pippin and Codlin, by contrast, were said not to thrive there.[24]

Orchards were also the scene of burials for those who were otherwise outside the established Church of England. In Sussex eighteenth-century nonconformists, such as Baptists, were frequently buried in orchards or yards when no appropriate meeting house and burial ground was available.[25]

In an uncanny reflection of events a century earlier in nearby East Grinstead, another Sussex orchard, this time at Chailey, was linked to a case of infanticide. The East Sussex coroners' records of 1812 stated that :

> [O]n 12 June Mary Tester, single woman, alone and secretly gave birth to a female child, a bastard, and was charged with assaulting her there and buried and concealed her with both hands. On 22 June the child was found in an orchard covered with mould or dirt…on that day Tester twice confessed in the surgeon's presence that she delivered herself of the child on 12 June, put her in a box and afterwards buried it in the garden or orchard. She declared that her child was not born alive and did not cry.

Tester, aged 21, pleaded not guilty and was acquitted of murder, but was convicted of secretly burying the child in an attempt to conceal the birth; she was sentenced to a fine of 1s and a year's imprisonment with hard labour at the Lewes House of Correction.[26] The poor infant was what country folk in Sussex called "chance-born" or a "love-begotten-child".[27]

For the orchards of Sussex, the decades from 1700 to 1840 proved to be something of a turning point. Responding to the growing interest from the gentlemen of the London Horticultural Society, increasing numbers of small Sussex nurseries adopted a more commercial approach to apple growing. The turnpike roads developed during the eighteenth century made it easier for nurserymen to send produce to customers, and around London such nurseries were well established by the 1750s, at Vauxhall and Putney, for instance. Nurseries also began to expand beyond London's suburbs, with skilled gardeners from great estates or more general gardeners setting up in business.[28] Young trees, shrubs, flower roots, bulbs and seeds were all commercial propositions. Carriers' wagons, carts and packhorses with wickerwork baskets of all kinds now plied the roads to the towns, laden with plants, seeds or young apple trees. As an indication of this growing interest, a list of more than 100 gardeners, florists, and nurserymen, mostly from Sussex, subscribed to the 1830 *Library of agricultural and horticultural knowledge*, including growers in Westergate, Yapton and Pound Hill and at Woods Nursery at Maresfield.[29] But growing apples in Sussex on a large scale for wider commercial markets had yet to come.

Endnotes for Chapter 5

1 Jane Austen to her sister Cassandra, 17 October, 1815 [http://www.janeausten.co.uk/magazine/page.ihtml?pid=276&step=4 – accessed 10 March 2011].

2 Morgan and Richards, *The new book of apples*, 66-7.

3 W.K. Ford, *Chichester diocesan surveys 1686 and 1724* Lewes: Sussex Record Society 1994, 173.

4 We are grateful to Christopher Whittick at the East Sussex Record Office (ESRO) for this reference. It is catalogued at ESRO AMS 5845/1.

5 West Sussex Record Office (herafter WSRO), Goodwood Mss E5531.

6 WSRO Petworth Mss 497.

7 The purchase of apple trees at Glynde Place 1754, source: ESRO, Glynde Mss 914.

8 G. Fussell, *More old English farming books: from Tull to the Board of Agriculture 1731 to 1793* London: Crosby Lockwood and Son Ltd 1950, 22-4.

9 Anon, 'Traditional Sussex cookery' *Sussex County Magazine* 10 (1936), 750. The English Folk Cookery Association was founded in the 1930s to promote regional recipes in England, but finished in 1940 with the death of its founder, Florence White, author of *Good Things in England* (1932). For a revival see http://www.englishfolkcookery.co.uk/index.html.

10 Rev. Arthur Young, *General view of the agriculture of the county of Sussex* (1813, reprint 1970), 162-3.

11 W. Marshall, *The rural economy of the southern counties* vol. II (1798), 192.

12 West Sussex Record Office PHA 90; 13142; 2234; MP 1752. We are grateful to Alison McCann and the head gardener at Petworth, Jonathan Arnold, for help with this enquiry; G.H. Kenyon, 'Petworth town and trades 1610-1760 Part I', *Sussex Archaeological Collections* 96 1958, 35-107; and Part II (1960), 109. Kenyon concluded that the probate inventories of Petworth innkeepeers provided "no more than the smell of cider and mead" since beer had become the staple drink in the area.

13 M. Reed, 'Social and economic relations in a Wealden community: Lodsworth 1780-1860' University of Sussex MA Dissertation 1982, 44-62.

14 Rev. Arthur Young, op. cit., 162. The Millennium parish map for Lodsworth depicts an orchard and notes that Lodsworth cider was renowned; Kim Leslie, *A Sense Of Place: West Sussex Parish Maps* Chichester: West Sussex County Council 2006, 15 2-3.

15 ESRO AMS 6448.

16 Quoted in Peter Brandon, *The discovery of Sussex* Chichester: Phillimore 2010, 43-4.

17 William Cobbett, *Rural Rides* (1st pub 1830; Penguin English Library 1967), 116.

18 Cobbett, op cit, 121.

19 Cobbett, op cit, 125-9.

20 Cobbett, 175.

21 Dr Carl Griffin kindly made his data base on plant maiming available, and see also C. Griffin, 'Protest practice and (tree) cultures of conflict: understanding the spaces of "tree maiming" in eighteenth- and early nineteenth-century England' *Transactions Institute of British Geographers* 33, 2008, 91–108.

22 British Parliamentary Papers *Report from the Select Committee on the Fresh Fruit Trade; together with the minutes of evidence, and appendix*.398 (1839).

23 J. Thirsk, 'Agricultural policy: public debate and legislation', in J. Thirsk (ed.), *The agrarian history of England and Wales V.II 1640-1750* Cambridge: Cambridge University Press 1984, 345-6.

24 *The library of agricultural and horticultural knowledge* Lewes: J. Baxter, 3rd edtn, 1834, 15.

25 J. Caffyn, *Sussex believers: Baptist marriage in the 17th and 18th centuries* Worthing: Churchman Publishing Ltd 1988, 68.

26 R.F. Hunnisett, *East Sussex coroners' records 1688-1838* (SRS vol 89, 2005), 166.

27 W. Holloway (of Rye), *A general dictionary of provincialisms* Lewes: Baxter and Sons 1838, 27.

28 B. Short, ' The South-East: Kent, Surrey and Sussex' in J. Thirsk (ed.), *The agrarian history of England and Wales V.I 1640-1750* Cambridge: Cambridge University Press 1984, 291; J. Harvey, *Early nurserymen* London: Phillimore 1974, 37.

29 *Baxter's library of agricultural and horticultural knowledge* Lewes: J. Baxter 1830, Appendix 9-11.

6. 1840–1914

Cock Wood.
183.
168.
169.
170
300
338.
299.
337.
302.
301.
304.
Woodside
298
297.
296.
340.
342.
343.
344.
336.
305.
303.
294
281.
348.
347
345
334.
349.
306
307
293
292
295
283.
308.
290
291
Cock Inn.
284
282
280.
350.
333
309.
289.
278
279
311.
285.
277
276.
352.
310.
288
287.
275.
259
332.
331
330.
Tanhouse farm
286.
274
260

6

Apples in Sussex: 1840-1914

"If apples bloom in March, in vain for 'em you'll sarch. If apples bloom in April, why, then they'll be plentiful; if apples bloom in May, you may eat 'um night and day."

East Sussex proverb, noted in 1883[1]

Pomologist at work: cataloguing and describing England's apple varieties was a central concern of leading members of the Royal Horticultural Society who gathered at a special Apple Congress in the society's Great Vinery at Chiswick in 1883

Facing page: Mapping orchards: this extract from an 1840 Sussex tithe map shows a substantial orchard in the grounds of a big house called Woodside in the parish of Peasmarsh (East Sussex Record Office, Peasmarsh TD/E12)

In Victorian times, the home-grown apple became the most popular dessert fruit for refined dining in grand country houses in Sussex. By the end of the century, the walled garden at West Dean, north of Chichester, for example, was, producing abundant fruit for the splendid dinner parties hosted by the American railroad magnate William James, owner of the estate. Today, the walled garden with its many orchard trees has been restored and stocked with many Sussex varieties such as Doctor Hogg, Crawley Reinette and Wadhurst Pippin, grown as open-crowned trees, with pears trained to grow around a variety of metal frames in Victorian style. The garden also boasts a

At West Dean Gardens, elegantly trained apple and pear trees thrive in an old walled garden, complete with thatched, flint-built apple store

flint and thatch apple store, designed to keep apples in good heart over winter.

Apples such as those at West Dean, grown on dwarfing rootstocks to keep trees small and bush-like, became popular in formal gardens throughout the county. Espaliers and wall-trained apples were fashionable too, producing both dessert and the more acidic cooking varieties. The distinction between the two roles is to some extent a matter of taste, and alters over time as the apples sweeten as they ripen. Traditionally, many apples were highly prized precisely because they were multipurpose – notably, in Sussex, the Forge apple from the East Grinstead, first described in 1851, although much older in origin. Forge was called "the cottager's friend" not least because it served equally for making cider, for cooking and for eating – it grew sweeter as it matured in storage, making it ideal for consuming fresh at Christmas.[2]

In the late 1830s and 1840s, much of England was painstakingly mapped, parish by parish, and specifically depicting orchards. These magnificent large-scale "tithe" surveys, recording landowners and land use, laid the groundwork for the abolition of tithes – archaic taxes which farmers were required to pay to tithe owners. Amounting to a tenth of all produce or profits, tithes were originally intended to support the upkeep of churches and the clergy. But by the 1840s, most tithe entitlements had been sold off to a motley collection of private landowers with no clerical connections. What's of interest here is that the maps show signs of orchards in about a fifth of Sussex's parishes; overall, out of some 300 parishes in Sussex, 65 had orchards depicted on the maps.[3] Tithe surveys covered holdings of all sizes, whether large farms or cottages, so they give a reasonably comprehensive picture of the number of orchards.

The middle years of the nineteenth century that followed the tithe surveys were something of a golden age for British farming, with good prices and decent crops. During this period, in 1854, the British

Pomological Society was formed to promote botanical and horticultural knowledge of apples. Joseph Paxton was the society's president; its secretary was Dr Robert Hogg (1818-1897), who became the leading authority on fruit of the Victorian period. When the Horticultural Society, jolted into action, set up its own Fruit Committee in 1858, Hogg became its secretary from 1860, and subsequently spent many years collecting and cataloguing apple varieties and advising fruit growers.

Robert Hogg was the son of a Scottish nurseryman. He studied medicine but became interested in botany, and joined a Brentford nursery firm, travelling widely throughout Britain and the continent to build up his knowledge of fruit and flower growing. Following his marriage in 1844 to Caroline Milligan (1821–1905), daughter of a Sussex corn factor from Worth, he became a co-owner of the famous Brompton nursery at Kensington, and when that partnership was dissolved in 1851 he went into the corn-dealing business with his father-in-law. That year he published *British Pomology* which featured 401 varieties of apple and was intended as a guide to the best apples to grow in English conditions.[4] In 1855 he became owner and editor of *Cottage Gardener*, later named the *Journal of Horticulture*, and in 1860 published the first edition of his *Fruit Manual*, a practical guide for nurserymen. By its final edition in 1884, more than 700 varieties of fruit were documented,[5] and it became the standard reference for English apple varieties. In 1861 Hogg took a law degree and was henceforth Doctor Hogg, or simply "the Doctor". In 1875 he became secretary of the Royal Horticulture Society, holding the position until 1884. In that year the great Apple Congress held at Chiswick received 1,545 varieties of apple, sent in to be classified and admired. His *Herefordshire Pomona* appeared in 1878, and was completed in 1885, documenting the many apples and orchards in the county.

Dr Robert Hogg (1818-97) revolutionised professional apple-growing by rationalising apple names and promoting the modernisation of orchards. An apple raised in Sussex is named Dr Hogg in his honour

In 1880 an apple raised at Leonardslee, Horsham, was named Dr Hogg in his honour. He was seen as an adopted son of Sussex, especially since he spent his leisure time on his farm at Heathfield.[6] At his death in 1897, the Royal Horticultural Society's instituted the Hogg medal, the society's highest award for fruit exhibits.

Hogg's career may have begun in a golden age of British farming, but by the 1870s a long downturn had begun, as poor seasons hit home production and foreign imports of grain undercut farmers' prices. During the 1870s imports from the United States and Canada in particular boomed as the American Civil War ended and transport improved. By 1900 these North American imports were joined by apples from Australia, New Zealand and South Africa, all popular with greengrocers, fruiterers and vegetable dealers, the fastest expanding retail trades in many towns around the county.[7]

At the RHS, Hogg and his colleagues began to argue that English apple varieties should be preserved and made more widely known. At the same time, Hogg spent much of his time persuading commercial growers to concentrate on a few "good" varieties. To compete with foreign imports, Hogg argued, English growers needed to specialise in a smaller number of popular but commercial varieties that were easy to grow in quantity and travelled well. The Ribston, the Blenheim and particularly the new Cox's Orange Pippin were continually compared and discussed, with a backward nostalgic glance at the older Golden Pippins.[8] As horticulturists like Hogg stressed the need to improve and "modernise" orchard management, as well as apple harvest and storage, older, regional varieties began to disappear from growers' catalogues.

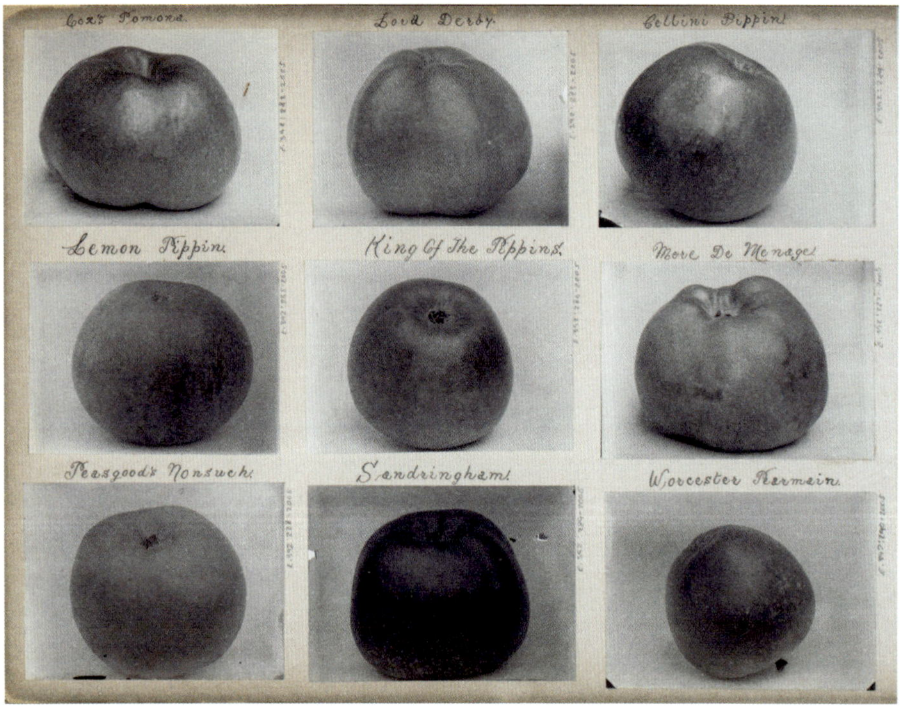

Charles Jones' photographic studies of fruit and vegetables, including his own apples, are now in the Victoria and Albert Museum's collection.

And while old varieties were becoming increasingly scarce out in the nation's orchards, the cultural significance of rare and exclusive apples among the wealthier classes tended to rise, as demonstrated by the photography of Charles Jones (1866-1959), both a professional gardener and an amateur photographer. His studies of fruit, including photographs of his own apple varieties, can be seen on the V&A website. Living on the Ote Hall estate, Burgess Hill, during the 1890s, he is best known today for his botanical images. The apples in these photographs are not Sussex varieties, for Jones was putting his more unusual varieties on show.

All the same, it was a good time for fruit-growing. As the agricultural depression continued, many arable farmers turned to dairying, while others diversified into apples. Already by 1883 the agricultural expert Charles Whitehead could look back on some years of definite progress in fruit farming in southeast England. In 1873 the national orchard area had been 141,000 acres but by 1904 this had risen to 236,000 acres, and to 245,657 by 1907. Some of this expansion was probably accounted for by improvements in government statistics, but equally holdings of less than one acre were excluded from the government figures.[9] Overall, between 1875 and 1895 the recorded area of orchards in England and Wales increased by 40 per cent and by 1900 the value of fruit and vegetable production exceeded that of wheat.[10] By 1914, fruit and vegetables came to account for perhaps as much as a quarter of total output of all other crops.[11]

Victorian gardener and amateur photographer, Charles Jones grew an impressive array of fruit, including this fan-shaped espalier pear, at Ote Hall near Burgess Hill.

During this optimistic time in English orcharding, horticultural education specificially designed for women began to develop, with fruit growing very much on the curriculum. New careers for women opened up, as a variety of training institutes were established, including, in Sussex, the Glynde College for Lady Gardeners. Launched by Viscountess Frances Wolseley (1875-1936) in 1905, the college lay on the chalk slopes below Mount Caburn around a house called Ragged Lands. Over the next decade, it trained scores of young women, many of them officers' daughters, as professional gardeners and growers. "They will go away, after their two years here, stronger, more capable, happier women, and out in the world they will be able in all honesty to say that much of this they owe to the teaching they have had, both in play and work time, under the chalk slope beneath one of the beacons of our English Downs," wrote Wolseley in 1916 in her glowing account of the enterprise, *In a College Garden*, published by John Murray.

Helped by men employed at the local chalkpits, the uniformed students laid out garden, walks and terraces planted with apples, pears and plums, with income from sales of the produce of the market garden, flower borders and orchard, all sent by train from Glynde to London. Frances Wolseley noted: "As I watch the young women gathering apples on this still autumn day, coming up the hill bearing on their arms wooden Sussex 'trugs', all lined with soft wood wool to prevent their load of fruit being bruised, I am more than ever convinced that gardening is most essentially suitable work for them."

She went on to describe the storage arrangements for the freshly picked fruit:

> Two long, low, dark and cool apple-rooms have been excavated out of the chalk, and here on wooden shelves the apples are carefully laid, forming thus the richest treasure-houses of the garden. Plenty of ventilation is provided, for we have a good succession of keeping kinds, and are able thus to reserve a supply for private customers right through the winter months.

The college was nothing if not highly organised, and each autumn the apple crop was closely monitored:

> A head student has charge of this Department, and whenever a fresh variety is gathered, she weighs the number of pounds with her set of scales, and enters it in a book. This record forms a valuable guide at the end of the season as to the amount of fruit the garden yields, and tells us, too, which kinds do best on this land. We have received recently the great encouragement of having some of the fruit grown here shown at the dinner which is now a yearly institution at the Mansion House, when a large and charmingly arranged cornucopia of produce is presented by the Worshipful Company of Fruiterers to the Lord Mayor. Amongst the apples that were sent by us as a gift to the Fruiterers' Company was one that weighed a pound and a half.

The Victorian and Edwardian interest in collecting information stretched to farming matters, and by the late nineteenth century, gives some indication of the growing popularity of orchards in Sussex across the county. In 1866 the Board of Trade (later successively the Board of Agriculture, and later the Ministry of Agriculture) began to collect annual returns from farmers relating to their crops and livestock. From 1873 the data, known as the 4th June Returns from the census date on which farmers completed the returns, included a section on orchard acreages.

Unfortunately, the data on orchards are patchy, with missing years and unexplained gaps for particular parishes, and are biased towards larger orchards – very small holdings were excluded.[12] Even so, using the data available, the distributions of orchards greater than one acre in area can be charted. In the maps that follow, the shadings show the percentage of orchard area set against the total area of crops and grass in each parish, so that the relative importance of orchards in parishes of different sizes can be compared.

In 1873 the largest orchard acreages, as a percentage of agricultural land devoted to crops or grass, were in the High Weald, in such places as Rotherfield, Ticehurst, Wadhurst and near Horsham at Lower Beeding, and also at Linchmere and Ambersham

The distribution of orchards in Sussex 1873

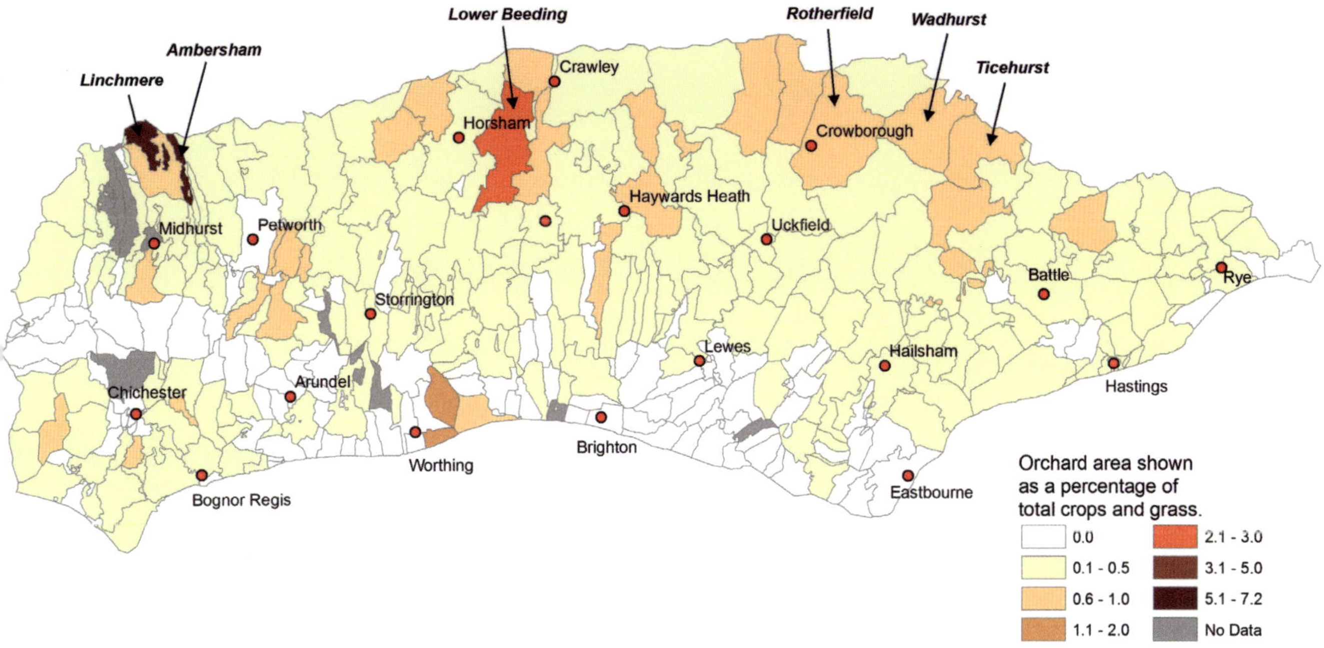

The distribution of orchards in Sussex 1907

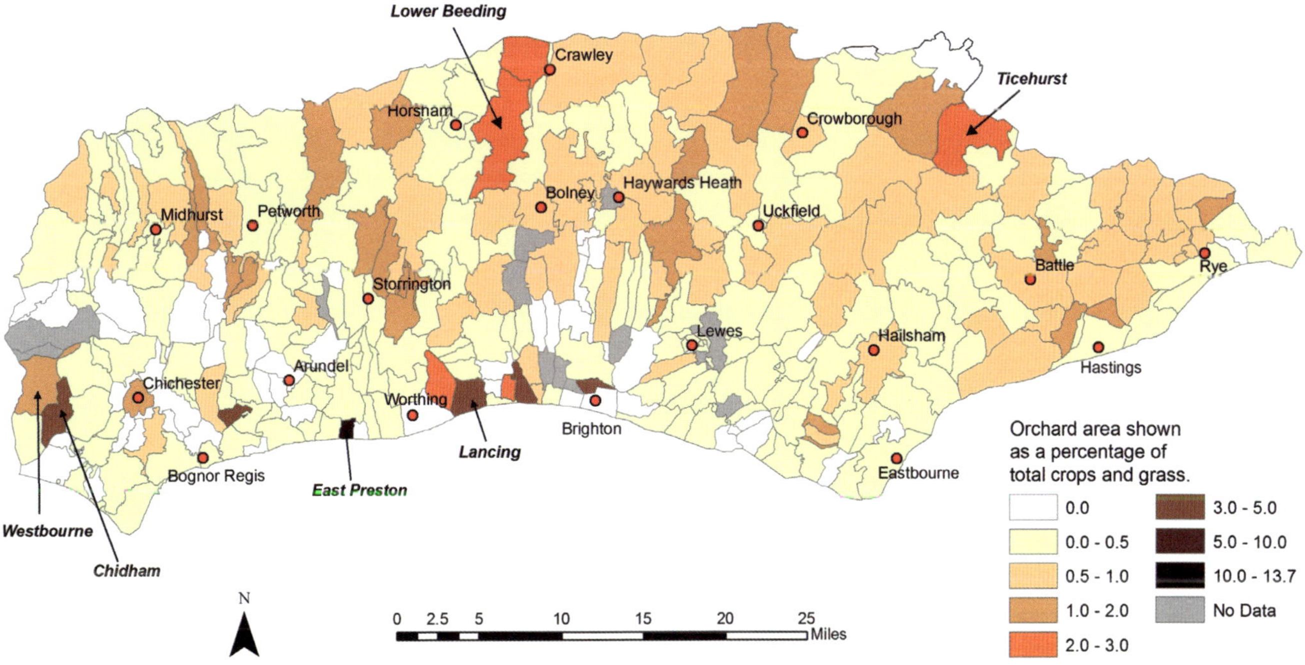

in the far north-west of the county. By contrast orchards were sparser in the chalkland parishes of the South Downs.

By 1907 high concentrations of orchards at Ticehurst and Lower Beeding were joined by orchards on the Sussex coastal plain at places such as Chidham, Westbourne, East Preston and Lancing. The total farm area devoted to apple orchards in Sussex, as opposed to other top fruit, was 2,294 acres – 68 per cent of the total land area dedicated to orchards in the county. So overall, between 1873 and 1907 the density of orchards in the county grew, and in the most favoured parishes doubled from around of 5-7 per cent to a maximum of 10-13 per cent of the total land area devoted to crops and grass. By 1907 orchards seems to have reached a new pre-First World War maximum, at 3,376 acres.

Alongside the burgeoning commercial enterprises, orchards were also being planted as an adjunct to the many large houses and gardens in Sussex. Thus at the "miniature villa residence" of Mill Hall, on the outskirts of Cuckfield, there is a report in 1879 of pleasure grounds and an orchard, with 1,200 apple and pear trees "planted under direction of Messrs Mitchell of Piltdown Nurseries". At Horsted House in Horsted Keynes in 1884 an estate agent's advertisement specifically noted the "well-stocked apple and pear orchards."[13]

However, not everyone was so approving. When the *Gardener's Magazine* Special Commissioner toured "the wasted orchards of England" (and published his findings in 1897),[14] he noted Sussex farmers' "want of energy, but equally the throwing together of farms leading to the neglect of orchards". He thought Sussex farms were becoming either too big to bother with orchards, or too poor to be able to invest in them. He noted the recent planting of orchards near Crowborough, but the trees were, he thought, poor, located on land sloping to the north east, lacking shelter, and with the trees planted too far apart to shelter each other. From here westwards to East Grinstead there were no orchards worth mentioning:

> The trees for the most part presented a miserably dilapidated and worn-out appearance, as if no care had been bestowed on them. This was more or less the case all the way to Three Bridges, and I could tell of many enclosures containing trees that presented such a starved decrepid appearance as not to have the redeeming feature of a touch of picturesqueness about them, but I hesitate to weary your readers with the depressing details.

Similarly, in 1910 agriculturalists Hall and Russell observed disparagingly that in many parts of Sussex orchards were small, cider-making rare and true cider apples not to be found.[15] Many Wealden farms in particular possessed small orchards that had become increasingly neglected as cider fell in popularity, following the national trend. In Sussex, as in Kent, orchards frequently competed with hop gardens for space on the farm, since the growing of hops for beer-making was a more profitable, if riskier, undertaking which could bring some wealth to farmers in a good year. Even so, the acreage devoted to orchards in the High Weald, which embraced both Sussex and Kent, rose from about 1,500 acres in the 1870s to just over 2,500 acres by 1890, rising slowly, with fluctuations, to over 4,300 acres by the outbreak of the First World War.[16]

And around Horsham commentators remarked on extensive orchards, though situated such a distance from a railway station as to make them difficult to access. Visiting the area on his pilgrimage around England to ascertain the state of farming in 1901, Rider Haggard estimated there to be 30,000 young fruit trees around Horsham, the largest concentration in Sussex at that time.[17]

Fruit tree nurseries in Sussex

One enterprise being set up during this period was that of J. Cheal and Sons near Crawley. The firm was founded by a Quaker family and became a leading tree nursery, also specialising in training fruit as cordons and espaliers, the use of topiary, and the growing of dahlias. An article in *The Garden* in November 1890 sang the praises of the firm:

> Crawley is famous for its fruit, and this reputation has been gained from the nurseries of Messrs. Cheal and Sons, who for the past twenty years, have paid great attention to the growth of apples, pears, plums, cherries, peaches and nectarines. The Lowfield Nurseries are about eighty acres in extent, but the acreage is constantly extending through the acquirement of surrounding fields. The situation is flat and low-lying, at one time a swamp, but now converted into a fruit garden by thorough draining and working the soil… a heavy loam.[18]

By 1891 their Lowfield nursery grew more than 70 different varieties of apple together with ornamental trees and dahlias.[19] Apple trees were available for growers and amateurs alike, with a wired fruit garden on display. The trees were planted as upright cordons, in double rows with three feet between each, and covered with wire. The wire was removed for picking, and replaced before flowering in early spring. In 1890 the varieties grown in this way were Cox's Orange Pippin, Loddington Seedling and the Yellow Ingestrie. Best of all, though, were the local Forge apples, but there were many other varieties to be seen, "cultivated with a careful and liberal hand". Joseph Cheal published *Practical fruit-growing* in 1893.[20] In 1906, the firm introduced to the market a popular Sussex variety, Crawley Beauty.

Cheal's also designed the gardens at Oxted Place in Surrey. The firm's drawings show that the gardens today remain very much as Cheal's designed them.[21] By the early twentieth century the firm was well known for its garden designs, often working in the Arts and Crafts style. In 1910 Cheal's created the Top Terrace and summerhouse at Standen in a similar style to Webb's earlier work, as well as the gardens at Polesden Lacey and Hever Castle. In the 1920s the firm also worked on the gardens at Woodmancote Place, near Henfield. The nursery was in business up until the 1960s though today much of its former site is covered by Gatwick airport and industrial estates.

The firm is probably best remembered for the apple Crawley Beauty as well as its crab apple *Cheal's Scarlet* – a small, crimson ornamental – a weeping cherry and, more prosaically, Cheal's roundabout in the centre of Gossops Green, Crawley.

Barnham Nursery is another good example of Sussex enterprise. The market gardening firm covering 120 acres was set up by the Marshall family from Brighton in 1881.[22] The census of that year shows that the two brothers, Henry Richard Marshall (aged 21) and Sidney Skelton Marshall (aged 18), were then trading together as nurserymen, employing 13 men and one boy. By 1895 the brothers had established separate businesses: Henry continued as a wholesale fruit grower and florist, with premises at both Brighton and Barnham, while Sidney established himself as S.S. Marshall Ltd. at Barnham, working as a nurseryman, landscape and market gardener, and fruit grower. Between 1905 and 1907 this firm became known as the Barnham Nurseries, and operating from a base opposite the railway station, ran large nurseries and fruit farms in Barnham and Eastergate – supplying London and provincial markets as well as the local area. After the war, the business gradually contracted, from 250 acres in 1948 to 100 acres by 1958. The Home Nurseries, by the station, were closed in the early 1970s, and the Old Nursery, opposite the Murrell Arms, was closed in 1981.[23]

At East Preston, George and Nelson Barnett established the first market garden in the parish in 1850 at "The Graperies", with apple orchards as well as grapes and flowers in greenhouses. The enterprise was taken over by Hubert Gray in 1892 and survived to 1934.[24]

Representations and orchard nostalgia

The Arts and Crafts movement in Victorian England encouraged the planting of traditional orchards and nostalgic cottage gardens stocked with old English flowers and fruit trees. The gardens of Gertrude Jekyll reflected these ideals, and she developed a close working relationship with William Robinson who moved to Gravetye Manor, West Hoathly, in 1884, and lived there until he died in 1935. Apple trees garlanded with climbing roses were grown for sight and scent among grass with snowdrops, narcissi and crocuses. At Gravetye Robinson was energetically planting trees including, in 1897, "the best cider apples known to French growers".[25] Standen, near East Grinstead, the 1890s house by Philip Webb, shows the Arts and Crafts inspiration stretching from the William Morris wallpaper to the walled fruit and vegetable garden, partly inspired again by William Robinson. The gardens still contain venerable Queen and Bramley trees, the Queen, widely planted in nineteenth-century gardens as a mid-season to late cooking apple, now showing signs of decay, the Bramley with its heavy branches propped up.

The depiction of flowers, fruit or vegetables became popular subjects for artists too. Still lifes of apples and paintings of apple blossom were popular among amateur artists and professionals alike. Such images featured heavily in the nineteenth-century vogue for rustic and seasonal settings, with autumn and apple harvest often linked together – in the romantic garden and orchard representations of Millais and Rossetti, for instance, as well as those of Miles Birket Foster and Helen Allingham, who came to live in Witley, near Haslemere. Huish, in his 1903 appreciation of Allingham, wrote with aesthetic understanding but with a sharp eye to the tree's worth. It was:

> ... one of those subjects which artists like to attack, a misshapen tree presenting every imaginable contortion... specimens of which are too often to be found in old orchards of this size, whose bearing time has long departed, and who now only cumber the ground, and with their many fellows have had much to do with the gradual decay of the English apple industry.[26]

The contortions referred to can be seen in Helen Allingham's *The apple orchard*.

Helen Allingham painted this view of a twisted, aged apple tree in around 1877, entitled The apple orchard

Henry Herbert La Thangue (1859-1929) studied at the Royal Academy in London and the École des Beaux-Arts in Paris. Disillusioned by the Academy, he became one of the leading figures in introducing French-style plein-air painting to Britain. He lived in Provence during his later years but maintained a base at Haylands, Graffham. He moved to Bosham, where he continued to paint large, social-realist works redolent of a Hardy-esque rurality, popular at a time of urbanisation and social change. He quickly attracted a wealthy clientele who responded to such scenes, showing contented

workers surrounded by rural calm and plenty. His first great Sussex work of this kind was the monumental A *Sussex cider press* shown at the Royal Academy in 1898. Others included *In the orchards: Graffham*, and the surrounding orchards and rural landscapes of Lavington and Petworth gave continual inspiration.[27] Harvesting the apples and storing them to await the arrival of a travelling press were labour-intensive activities often involving wives and children during the weeks of autumn, and cider-making lasted, in some instances, through to December. La Thangue carefully watched the whole process, from gathering wind-falls to loading laden baskets on to carts.

Rural idyll with child labour: Cider apples, by Henry Herbert La Thangue. His images of rural calm and plenty were somewhat at odds with the reality of the day

Unlike their pale city counterparts, La Thangue's yeomen apple gatherers were depicted as strong and healthy, with women and men, young and old working together. In this artist's hands, apple picking became a metaphor for rural harmony in the face of industrialisation.[28] In fact, such an idyllic picture was rarely true. Pickers were poor, frequently unhealthy, and had often travelled long distances. Local women and children were employed, together with itinerant labourers, working in gangs, and paid in part with cider. The cider was frequently consumed in the fields during harvest times, and larger farms in West Sussex maintained orchards expressly for this purpose.

The Victorian and Edwardian period witnessed a growing interest in fruit-growing in Sussex, matching the burgeoning of scientific research within the Royal Horticultural Society. Larger acreages were planted, and apples and other fruit were now being transported more efficiently to London and other urban markets, where demand for fruit was increasing. And the orchard, the space which was both "natural" yet "homely", increasingly became the setting for depictions of the idealised rural idyll.

A Sussex orchard, painted by Henry Herbert La Thangue in 1905

Endnotes for Chapter 6

1 G.L. Apperson, M. Manser, and S. Curtis, *Dictionary of Proverbs* London: Wordsworth Editions Ltd 2006, 23.

2 J. Morgan and A. Richards, *The new book of apples* London: Ebury Press (rev edtn 2002), 89.

3 R. Kain and R Oliver, *The tithe maps of England and Wales* CUP 1995, 517.

4 R. Hogg, *British Pomology; or a History, description, classification and synonymes of the Fruits and Fruit Trees of Great Britain. [vol. I.] The Apple*. London 1851.

5 R. Hogg, The Fruit Manual; containing the descriptions & synonymes of the fruits and fruit trees commonly met with in the Gardens ... of Great Britain, etc London 1860. For a full discussion of Dr Hogg and the activities of the apple enthusiasts at this time see B. Elliott, 'Hogg's Fruit Manual, its rivals and successors: a bibliographic study', *Occasional Papers from the RHS Lindley Library* Vol. 4

(October 2010), 5-36.

6 See Chapter 6 for details of the Dr Hogg Sussex variety. For his life, see M. C. Curthoys, 'Hogg, Robert (1818–1897)', *Oxford Dictionary of National Biography*, Oxford University Press, May 2009; online edn, May 2010 [http://www.oxforddnb.com/view/article/96781, accessed 15 March 2011]; and see also B. Elliott, 'Fruitful endeavour', *The Garden* October 2010, 694-7.

7 B. Afton and M. Turner, 'The statistical base of agricultural performance in England and Wales, 1850-1914 in E. Collins (ed.), *The agrarian history of England and Wales VII 1850-1914 (Part II)* Cambridge: Cambridge University Press 2000, 2115; M. Winstanley, *The shopkeepers' world 1830-1914* Manchester: Manchester University Press 1983, 152-66.

8 Morgan and Richards, *The new book of apples*, 93-4.

9 Report of the Departmental Committee upon the Fruit Industry of Great Britain (*British Parliamentary Papers* 1905 [cd 2589], 2.

10 C. Whitehead, 'The progress of fruit farming' *Journal of the Royal Agricultural Society* 19 (1883), and reproduced as *The progress of fruit farming* London: W. Clowes & Sons 1885; and 'Fifty years of fruit farming', *Journal of the Royal Agricultural Society* 25 (1889), 56-79.

11 M. Turner, 'Agricultural output, income and productivity' in Collins (ed.), *The agrarian history of England and Wales VII (Part I)*, 286-7.

12 The Sussex data for 1907 is at The National Archives, MAF 68/2257.

13 British Library, Sale Catalogue, Maps 137 b.10.

14 Gardener's Magazine Special Commissioner, *The wasted orchards of England: some truths about British fruit growing* (W., H, and L. Collingridge 1897).

15 B. Short, Agriculture in the High Weald of Kent and Sussex 1850-1953 (PhD thesis, Univ. of London 1973), 170; Hall and Russell, *Agriculture and soils of Kent, Surrey and Sussex*, 37.

16 W.E. Bear, 'Flower and fruit farming in England' pt 3 *Journal of the Royal Agricultural Society* 10 (1899), 63.

17 H. Rider Haggard, *Rural England, being an account of agricultural and social researches carried out in the years 1901 and 1902* London: Longmans 1902, 130.

18 Anon., 'Fruit at Crawley' *The Gardener* 1 Nov. 1890, 419.

19 A.M. Benton, *Cheals of Crawley* Moira publications 2002. Alice Benton was the granddaughter of Ernest Cheal (d. 1961), son of Joseph Cheal.

20 J. Cheal, *Practical fruit growing* London: G. Bell and sons 1893.

21 West Sussex Record Office Add Mss 28,700 — 28,718 1834-1959 The records relate primarily to the business of John Cheal and Sons, nurserymen, seed merchants, florists, and landscape gardeners, at Lowfield Nurseries in Crawley and Charlwood on the Sussex-Surrey border.

22 A short notice, probably written by the Company itself, gives the date of establishment as 1879 (*Agriculture in Sussex*, The County Associations Ltd 1948, 35).

23 See S.S. Marshall accounts journal, WSRO Add Mss 41,252 29 September 1898 — 30 September 1903.

24 *VCH* 5 (ii), 260.

25 W. Robinson, *Gravetye Manor; or, Twenty years' work round an old manor house. Being an abstract from the tree and garden book of Gravetye Manor, Sussex, kept by the owner, W. Robinson* London: John Murray 1911.

26 Marcus B Huish, *The happy England of Helen Allingham* London: Bracken books 1985 (orig pub as *Happy England as painted by Helen Allingham* 1903), plate 69 and p. 167.

27 Kenneth McConkey, 'La Thangue, Henry Herbert (1859–1929)', *Oxford Dictionary of National Biography*, online edn, Oxford University Press, Sept 2004 [http://www.oxforddnb.com/view/article/64497, accessed 8 June 2010]

28 Oldham Art Gallery, *A painter's harvest: H.H. La Thangue 1859-1929* Oldham Libraries, Art Galleries and Museums 1978.

7. 1914-2012

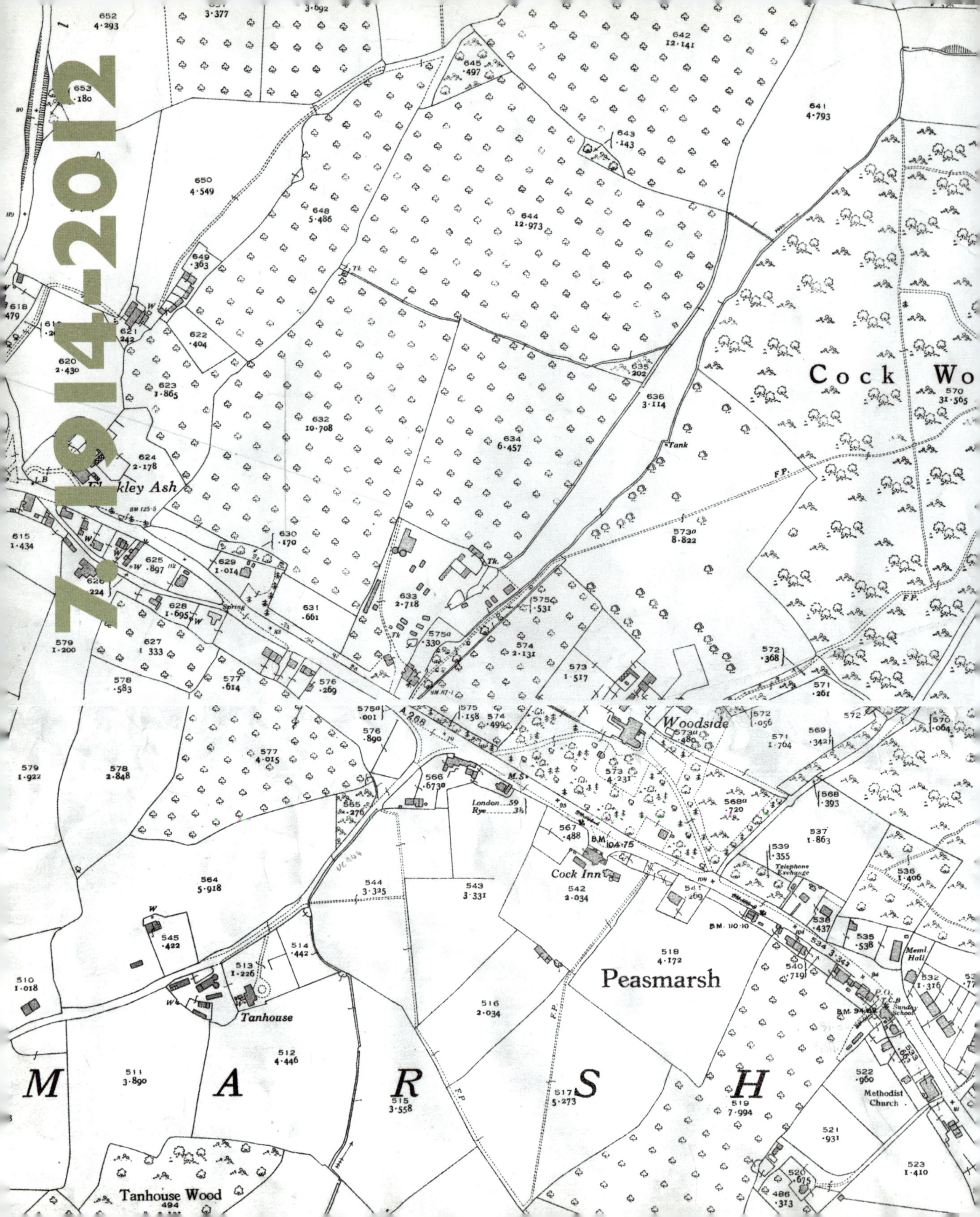

Cock Wo
Woodside
Cock Inn
Peasmarsh
Tanhouse
Tanhouse Wood
M A R S H
Methodist Church
Meml Hall
Telephone Exchange
Sunday School
London 59
Rye 3½
A 268
Tank
F.P.

7

Apples in Sussex: 1914-2012

> *"We idly grumbled at the yield*
> *Of apple-orchards in the Weald"*
> *Vita Sackville West, 1921*[1]

In the early decades of the twentieth century, orchards in Sussex were burgeoning enterprises. Across the county, more and more land was being planted with fruit trees, especially apples. On the coastal plain, commercial fruit farms and nurseries thrived. In parts of the Weald, fruit growing was seen as something of a saviour in difficult times for farming, and became a major focus of many farms, especially at Wadhurst, Ticehurst and Salehurst in the northeast near the Kent border, and around Kirdford in the west. Elsewhere in the Weald, many landholdings had smaller orchards that were valued particularly for their complementary role within mixed farms.

At Bolney, for instance, apple orchards were planted in the 1930s as part of a dairy farm. Fruit was sold at local shops as well as through regional wholesalers, and food waste was minimised, because all manner of produce could find willing consumers – even scruffy little apples could be sold cheaply in local greengrocers and windfalls processed for juice or cider, or fed to livestock.

Glyn Stephens took over the land at Bolney in the 1980s, but stopped trading in apples in the early 90s when he found he was losing money on the fruit. But the orchard persisted and today, Old Mill Fruit Farm, with 40 acres of orchard and 8500 trees, is being refurbished by his son Tom, who has started producing juice and cider, under the distinctive brandname Wobblegate, and sold mostly locally to pubs and restaurants. The orchard isn't certified organic but the trees haven't been sprayed since 1992.

Facing page: Orchard boom - a century on from the 1840 tithe map in chapter 6 (p86) which indicated lines of fruit trees in the grounds of Woodside in Peasmarsh, the parish is now awash with orchards as shown in these Ordnance Survey maps of 1939

At the 1400-acre Broadhurst Farm at Horsted Keynes, John Clarke planted 170 acres of the best-drained land on the estate with fruit trees during the depression of the 1930s. His son, Rory Clarke, remembers that his father "always liked to think big". At part of the Ashdown Land Company, John planted 20 acres with raspberries and blackcurrants and put in a few peaches too. He stocked the plum orchards with Purple Pershores, Gages, Victoria and Oullins Golden Gage, and for the apples chose Cox, Worcester, Lord Lambourne, Bramley, Lanes Prince Albert,

George Cave and Egremont Russet. During the late 1930s, John Clarke built a packhouse and stores and started a canning factory for the soft fruit, which was marketed together with the apples and plums under the "Blue Smock" brand. Rory Clarke says that the Ashdown Seedling variety originated from the pip of a McIntosh apple that his father had eaten. The original tree grew in the kitchen garden at his home, and the pip is celebrated in the advertisement for the "Blue Smock" brand in 1932.

The remaining farmland at Broadhurst formed a mixed farm of dairying, sheep, corn and woodland. The two dairies on the farm milked a herd of 300 cattle and fattened up store cattle for beef. Sheep grazed the poor sandy and clay soils while a small area of 50 acres was cultivated for grain. Under the 400 acres of oak woodland there were hazel, hornbeam and chestnut which were coppiced for thatching spars, charcoal and stakes.

Such a mixed farm employed a large number of people, many of whom lived on the estate. Under the farm manager and farm foreman, there were two dairymen plus relief staff, a shepherd plus his assistant and three woodmen. In the orchards there were about 20 people with a manager, foreman and packhouse foreman. The orchard workers were employed all-year round, picking fruit from August, then packing until February and March. At other times they worked on the farm. A large seasonal workforce came down from London to help with the fruit picking, and many of these workers slept in old railway carriages on the farm.

These two examples, at Bolney and Horsted Keynes, were part of a sustained and widespread orcharding movement in Sussex during the early decades of the twentieth century. The 1920s and 1930s were a time of depression in farming, with arable farmers hard hit by falling prices. To survive financially, many farmers turned to alternative enterprises, particularly fruit growing. In fact, an impressive rise in fruit production in Sussex can be documented from the 1920s, thanks to the decision of the Ministry of Agriculture to expand the scope of its annual census of agricultural land use.

Health and tradition, in a tin: Ashdown Land Company began a canning enterprise at Horsted Keynes in 1932, and drafted in an image of a genial traditional "smock frock" farmer surrounded by orchard blossom to promote its produce

In 1922, the ministry collected returns from 172 fruit growers in Sussex, revealing particular concentrations of professional apple growers in four regions of the county: in central Sussex, including Newick and Lindfield; in the Washington, Thakeham, and Warminghurst area near Storrington; on the coastal plain around Worthing; and in the far west around Chidham and Westbourne, on the Hampshire border.

The following year, the ministry's census revealed more detail, listing the number of individual orchard trees per parish – as many as 12,000 in Sedlescombe and 5,721 in Battle, for example, adding up to a grand total of 617,071 orchard trees in Sussex! These data show high numbers of trees (per 100 acres / 40.5 hectares of crops and grass) at Salehurst and Sedlescombe, on the coastal plain at Worthing and in the far west at Westbourne (13,000 trees). A comparison of two maps, one showing the distribution of growers and the other the fruit trees themselves, suggests that orchard holdings were somewhat larger at Salehurst and Sedlescombe, compared to central-mid Sussex and around Storrington. The census also recorded the acreages of orchards with crops beneath and between the trees, such as red and black currants, raspberries or gooseberries, as well as those left fallow or with permanent grassland. These statistics show that it was not uncommon for apple growers to produce soft fruit as well within orchards – a practice of intercropping that bears some resemblance to the "forest gardening" favoured by today's permaculturists (see chapter 11).

Facing page: Eve and the apple: in the 1930s, the Ashdown Land Company also harnessed the allure of glamour to sell apples – a far cry from the twenty-first century 5-a-day campaign

Yet despite the growth in orchard acreage, a new threat loomed in the interwar years – an unprecedented burst of suburban development. Gardens and orchards around the towns of Sussex proved no impediment to the construction of both council and private houses. Near Brighton the orchards and market garden of Hodshrove Farm were engulfed by housing, in this case making way for the Bates Estate at Moulsecoomb, replacing the Bates Nursery, beloved of former generations of schoolchildren for its scrumping possibilities. A similar fate awaited Woollards Nursery, well known for its

Number of fruit growers per parish 1922

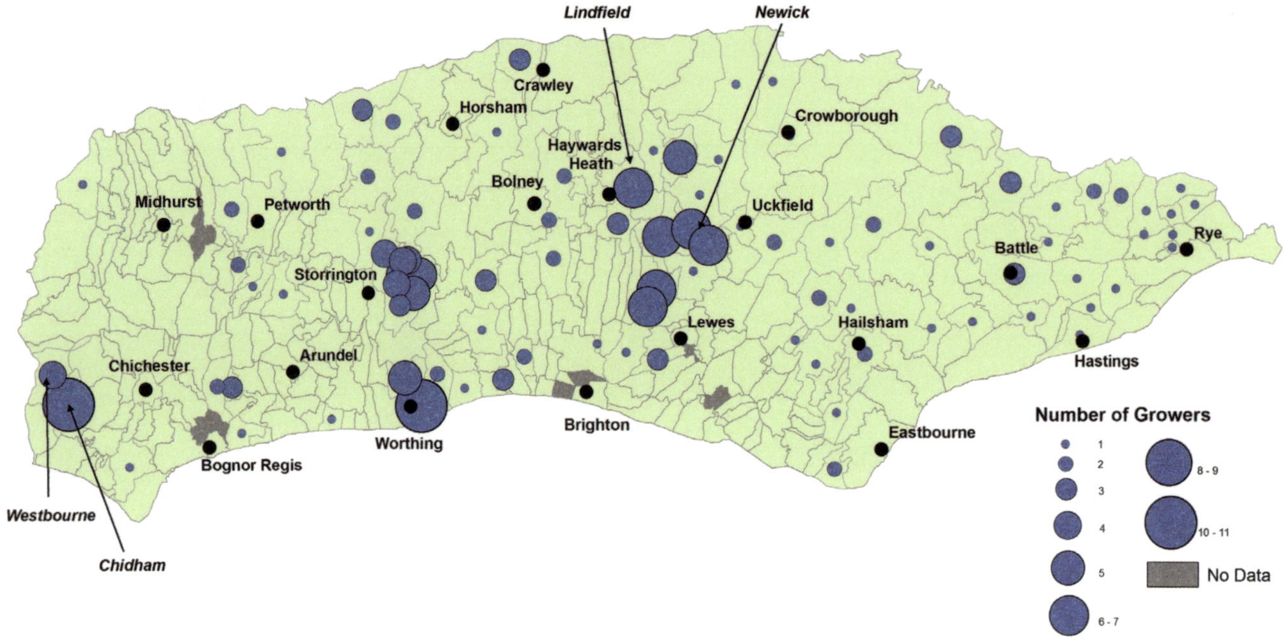

Number of orchard trees per acre of crops and grass 1923

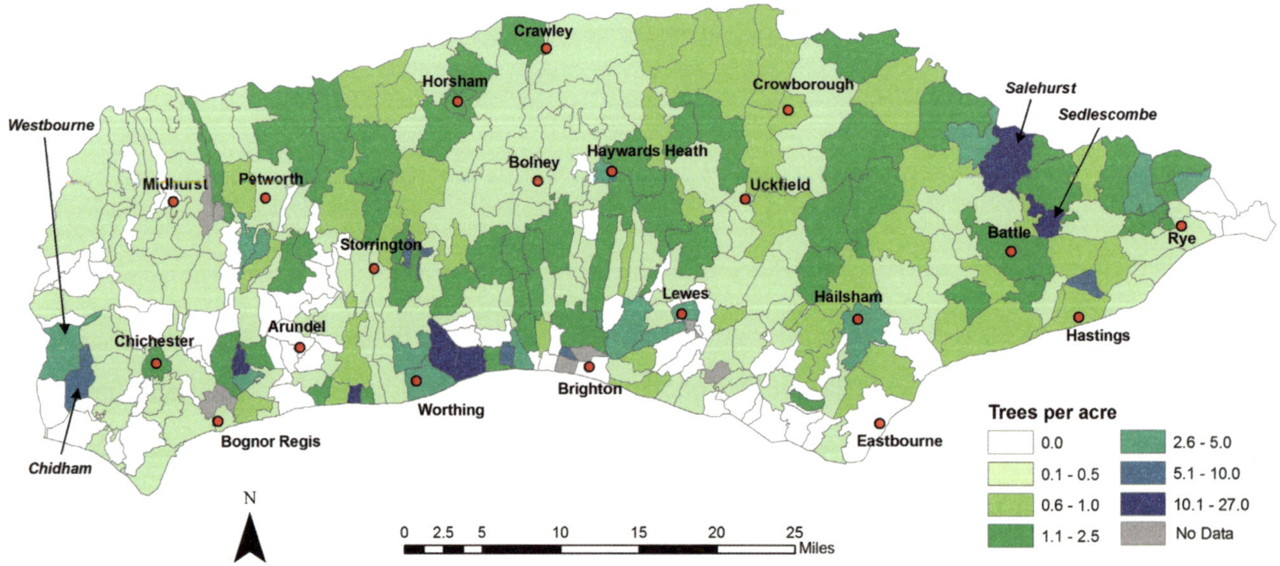

apple trees, which fronted the old road (now A27) west of Falmer which has now become part of the campus of Sussex University. The nursery served a greengrocer's shop in Brighton's Lewes Road. The attractive fruit orchards included apples, pears, cherries, and quince, as well as flower gardens. And one of the apple trees was an exotic White Transparent, a Russian variety that fruits unusually early, in August. The grass pathways that linked the flower gardens were themselves full of wild flowers.[2] Ruby Dunn in her memoir captures the scene:

> All the land on the west side of Lewes Road, between Moulsecoomb Place and Coldean Lane, was given over to fruit farms and nurseries, some of which belonged to Mr Bates and some to Mr Woollard. I remember the time when I was told that I was old enough to be allowed to cross Lewes Road on my own and walk to Bates Nursery to buy the fruit for our family; the scent of the stored fruit as I entered the wooden sheds, and watched the wind fall apples and pears piled high in the wooden bushel measure, is with me still. There were also damsons and plums to be selected for Mum's annual jam making and fruit bottling.[3]

Although the rescue of older regional varieties remained very much a minority interest, the National Fruit Trials, begun by the Royal Horticultural Society in 1936, did take up this project in an attempt to widen knowledge of the available varieties. One Sussex variety, Duck's Bill, was received for RHS trial in 1937 from Fred Streeter, "the radio gardener" and Head Gardener at Petworth House in Sussex. Streeter had been head gardener from 1911 to 1913 at Lavington Park, Petworth, the seat of the whisky magnate James Buchanan. Wounded in the First World War, Streeter was invited by the third Baron Leconfield to be head gardener at Petworth House in 1926. He remained there for the rest of his life, on the estate where his mother's parents had been tenants. His accomplishments included a sunken garden, the renovation of the kitchen garden which fed the lavish house parties at Petworth, and the growing of specialist fruit such as peaches, nectarines and quinces. He won more than fifty gold awards for exhibits at RHS shows. To many he was best known as a broadcaster: his career beginning in 1935 when he was invited to take part in the BBC radio gardening programme *In Your Garden*. His enthusiastic, rolling Sussex burr, brought 200 letters of appreciation and Lord Leconfield is quoted as saying to him afterwards: "only the aristocracy knows me, Streeter. But after today the world knows you."[4]

Radio personality: Fred Streeter brought a Sussex apple with one of the most intriguing names – Duck's Bill – to the attention of the wider world, and shared his gardening skills with a generation through weekly BBC broadcasts

Men like Fred Streeter who had survived the horrors of the First World War and who now sought to build a new life in peacetime provided another boost for orchards. General Sir Ivor Maxse (1862-1958), active in France during the war, retired from the army in 1926 and promptly began a second career as a commercial fruit grower from his home at Little Bognor, Fittleworth, which he shared with his wife Mary Wyndham (1870–1944), eldest daughter of the second Baron Leconfield. It seems to have been a progressive orchard for its day, for instance, making use of the then relatively new M9 dwarfing rootstock, and experimenting with peach production on a nearby terraced site in Little Bognor. Maxse sold apples such as Worcesters under the name of the Maxey Fruit Company. When he visited Rudyard Kipling at Batemans in September 1933, he offered his advice on the orchard there, apparently recommending its grubbing up.[5] There's still an orchard at Little Bognor House, producing mostly Cox and Golden Delicious, says gardener Sean Bowyer, and self-catering holidaymakers can stay in the handsomely converted "potting shed" where orchard workers once queued to receive their wages.

Handled with care: Worcester apples sorted by hand off the Maxey Fruit Company's rolling belt near Fittleworth 1959

In the 1930s, there were moves to improve the quality of British apples, as the government introduced the National Mark Scheme, setting out standards for luxury and cheaper grades and a national labelling scheme for apples and pears through accredited packers.[6] In 1932 free-trade policies were abandoned and import duties levied on non-Empire goods, such as those from the US and Europe. As a result, the competition on British apples was much reduced, and came mainly from Canadian apples. Those imported from Australia, New Zealand and South Africa did not clash with British producers because they were imported for sale only after British apples, available fresh in the autumn and then stored over the winter, came to an end. Apples from the southern hemisphere then arrived in greengrocers to fill the gap until British summer varieties became available again. Over the 1920 and 30s, a growing demand for better diets helped to enhance fruit sales. Overall, between 1914 and 1936 the consumption of vegetables rose by 64 per cent and fresh fruit by 88 per cent.[7]

The orchard area in Sussex increased in size by a quarter in the space of 16 years, with the greatest expansion at Kirdford near Petworth. The area bordering on the High Weald of Kent remained a pre-eminent orchard region, with a string of parishes between Ticehurst and Udimore rich in orchards as well. Small domestic orchards were also plentiful in the Weald, with poultry, pigs and sheep grazing among the trees. Larger orchards were often linked to nurseries selling fruit trees and garden plants and located near railways, such as the Cuckoo Line between Tunbridge Wells and Polegate, which didn't close until the 1960s. Complementing the development in Kirdford in western Sussex, an apple grading station was also established at Robertsbridge, in the far east of the county.

Dressed to impress: young women display carefully packed apples in what appears to be a rather manicured packing shed – preparation for a special visit perhaps?

The coastal plain was also home to clusters of orchards around Worthing and Westbourne, as well as substantial areas of nurseries and glasshouses. Near the Chichester to Portsmouth road and railway, acres of apple orchards flourished on the fertile brickearth.[8] By 1939 orchards had expanded from the Kent border parishes across Sussex. Even in the South Downs, the valleys with deeper soils and shelter from the wind were peppered with orchards – around Brighton, for instance,

The distribution of orchards in Sussex 1923

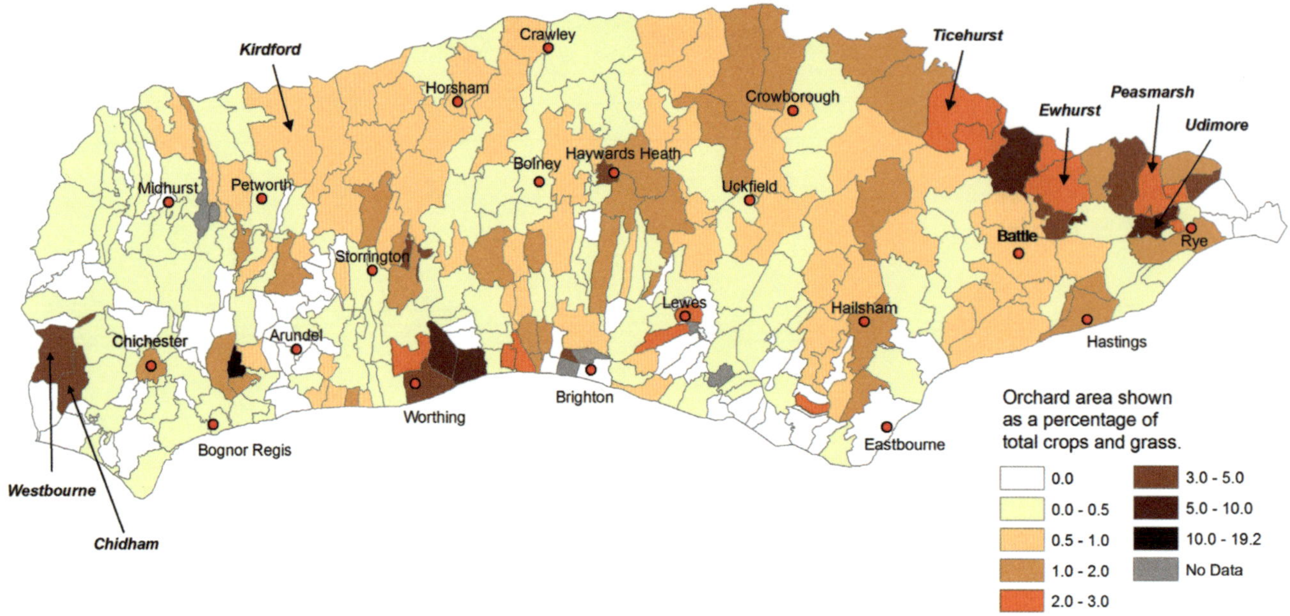

The distribution of orchards in Sussex 1939

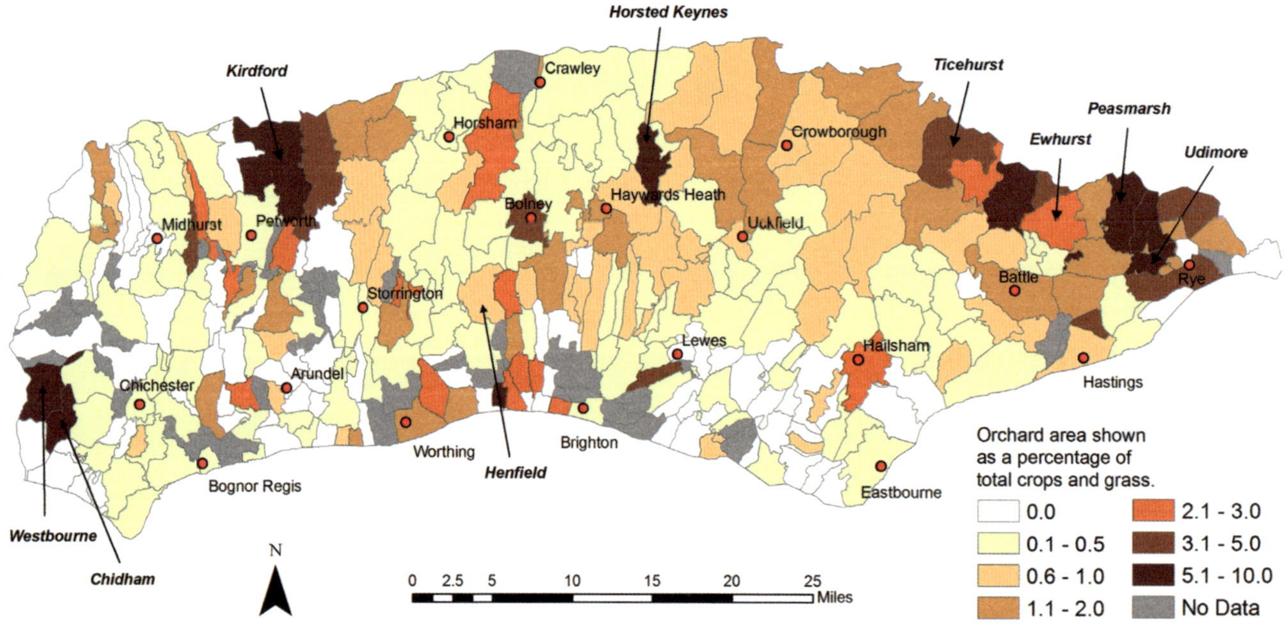

Sussex orchards prosper between the wars: selected parishes

Parish	1923 area (hectares)	1939 area (hectares)
Battle	14	28
Chidham	27	32
Ewhurst	25	41
Kirdford	14	168
Peasmarsh	34	104
Udimore	40	70
Westbourne	45	68
Sussex as a whole	2003	2528

Source: Ministry of Agriculture 4 June Returns

and at the foot of the downs at Kingston-near-Lewes. On the Lower Greensand, particularly between Ditchling westwards to the Arun, orchards and nurseries flourished, especially near Henfield.

The rise of Kirdford's orchard co-operative

The personal stories behind the vast increase in acreage recorded in Kirdford parish near Petworth are particularly noteworthy for the history of apple-growing in Sussex. In 1925 Bill Fowle purchased Costrongs Farm and was soon inspired to plant apple trees on the Weald Clay, an area marginal for cereals when prices fetched were low. From nearby Wisborough Green, A.J. Carter sent apples to Covent Garden where they had impressed Dr Hatton, director of East Malling Research Station, who suggested that the locality might be suited to specialist fruit production. A number of other incomers, returning from work as tea planters in the British colonies, for example, purchased nearby farms and turned much of their acreage over to orchards. The growers included Messrs Baker, Napier, Spence and Holdsworth, to be joined by Hugh Kenyon, also a noted Sussex local historian, in 1928. Imaginatively, a Kirdford Growers Friendly Society was formed, which soon became a fully-fledged cooperative, with 20 members at its height.

Tom Mursell of Dounhurst Farm, Kirdford, says his orchard joined the cooperative after his widowed grandmother moved from Cheshire to Dounhurst Farm, Wisborough Green, in 1932, and planted 120 acres (49ha) of apples and pears.[9] Bill Fowle planted Costrong with Cox's and Bramley, continuing with his orchards until he emigrated to New Zealand in the 1950s.[10]

A packing station was opened in 1934 by Tom's father, Sir Peter Mursell, with the other Kirdford farmers, providing work for many locals in the sorting, grading and packing of apples, as well as pears and cherries, all sent to London, as well as work in the large storage sheds and cold stores. Gladys Marshall, now in her late 80s, worked as a picker at Kirdford for some fifty years, from the Second World War onwards (see interviews, chapter 8). Reg Thompson, 86 when interviewed in 2010, was born in Kirdford, and remembers the days when many villagers were employed on the farms or at the packing station, and the heyday of the Kirdford Growers co-operative. "You couldn't beat a Kirdford apple," he said, "Apples were at the heart of Kirdford.... It was open to the public, there were demonstrations. There was a shop and even a cafe. All sorts of people came from miles around."[11] Briault, writing up the results of the 1930s Land Utililisation Survey for Sussex, remarked particularly that "high quality produce is found on the young plantations" at Kirdford.[12]

The Second World War and its aftermath

During the war, orchards in Sussex were inevitably somewhat neglected, with labour in short supply. But many producers managed to keep going, and indeed the overall index of prices for fruit, vegetables and glasshouse products rose throughout the war, as food imports effectively ceased.[13] On the fruit farms at Kingston-near-Lewes, a crop reporter working for the Ministry of Agriculture in 1941-43 noted that Major R. Everett had "a well managed fruit small holding. Two glass houses are used for tomatoes and most of the land is planted with apple trees, market garden crops being grown on about one acre." Peter Sharp was a conscientious objector during the war and in 1941 was working on the land at Ripley in Surrey. He answered an advert for a "compost gardener" at a small progressive school in Forest Row. Compost gardening was an early expression of the organic movement and its supporters "used no artificial". After two years he placed an advert in *The Grower* magazine as Grower Foreman and received five replies, accepting a job with Major Everett in Kingston. Peter, as a conscientious objector, was surprised to be employed by an ex-army man.

Major Everett's neighbour, F.R. Wettern, had an 11.5-acre orchard and a grass field, attached to his house. He had been living there for three years. The reporter noted: "Formerly the orchard had been badly neglected: the trees have [now] been pruned and regularly sprayed and now good crops of apples and plums are grown. Sheep from neighbouring farms are grazed in the orchard and in the grass field." At Kingston Ridge was Major Snoxell who ran "a very well managed fruit and market garden holding, apples, pears, plums and soft fruit being grown and all available space between the trees being cropped with market garden produce."[14]

Sussex orchards were frequently tended by Women's Land Army personnel. Joan Culver was sent from Plumpton College to one of the apple farms at Kirdford. She

lodged in the farm house and cycled to work at 7am, working with three men, all aged well over 50. Pruning was the main job in winter, but on wet days she would work in the barn cutting sacks into narrow strips to make grease bands for the trees. In spring spraying would begin.[15] The picture of Women's Land Army girls at work was conceived and sketched at East Malling Research Station in December 1944 by the war artist Evelyn Dunbar (1906-1960), who spent much of her life in Kent. The painting has a central panel surrounded by a frieze, and the viewing point is from the top of a ladder. On the right, a woman is pruning a tree, and beyond her, two more women reach into the trees they are pruning. Another woman carries a tripod ladder. In the distance, other land girls are pruning trees or collecting prunings from the ground. Grey winter light predominates, the women are all well wrapped up. More usually, women are depicted undertaking menial tasks, such as weeding or picking under the eye of a male foreman, but in this instance a team of women

War artist Evelyn Dunbar painted a powerful tribute to women's war work in the nation's orchards. Oil on canvas, A 1944 Pastoral: Land Girls Pruning at East Malling

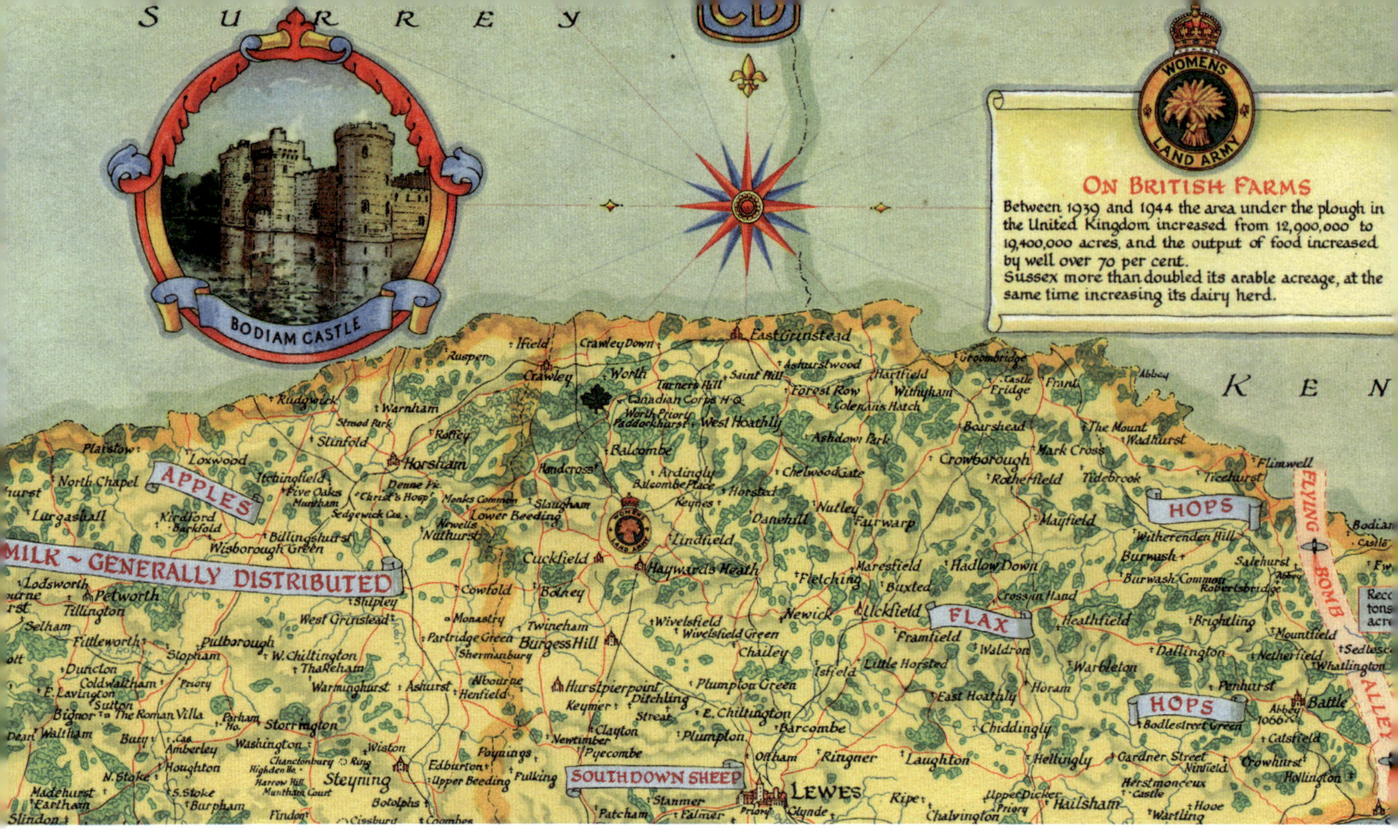

Orchards in the picture: another image from a war artist of this era, Ernest Clegg's Map of Sussex – one of a series of county maps drawn in1945 to commemorate the work of the Women's Land Army – indicates that apples were a significant crop in part of West Sussex

are carrying out skilled work in a time of war. The frieze portrays apples and gloved hands holding secateurs and saws.

During the war, the writer John Stewart Collis worked on the land. On a fruit farm in Sussex, his first job in April 1940 was to collect and burn the cut branches of apple trees. He wrote of spreading potash around the fruit trees, hoeing, harrowing between trees with a horse-hoe and spraying them. The spraying machine, mounted on the chassis of a Morris Oxford, had a tank filled with water, sulphur and lime "a combination of such a poisonous complexion that even to look at it was mentally disturbing … [and] seemed bent on spraying me personally with its unspeakable juices".[16]

After the war, the land devoted to orchards continued to increase, and by 1955 there were 2,884 hectares in Sussex, the highest area of orchard cultivation on record. Overall, the period from the 1870s to the mid-1950s saw a doubling of the overall area of land devoted to orchards in Sussex.

In the Weald of Sussex the orchards on many properties were an important part of both the landscape and the rural economy. For instance, on the Steep Park Estate, Rotherfield, an "Orchard Holding" of some 6 acres with a Cox's Orange orchard of 3½ acres was advertised for sale in 1955. Next door was a market garden with a former oast house converted into a grading shed and apple store, and a separate packing shed. At Holmes Farm, Hooe in 1951 a well-stocked apple orchard contained Bramleys, Lane's Prince Alberts, Newtons and James Grieve apple trees as well as Coxes, which filled half of the orchard area.[17]

Cider-making in Sussex still persisted after the war, though probably only in a few places. When travel writer Hardiman Scott toured the county in 1947-48, he set out to capture "the little-known Sussex – people and things which have been seldom or never recorded elsewhere in books." In his book, *Secret Sussex*, published in 1949, he described his visit to "cider-man" Jim Bicknell, who lived in "a little valley called Hillgrove Common, about a mile west of Northchapel."[18] According to Scott, Mr Bicknell's trade was really besom-making, crafting sought-after brooms from birch cut from the woods that rose steeply from his house. He only made cider to private order, and every autumn called upon his old customers to inquire how much they wanted. "He had been doing this almost every autumn of his adult life, and before him his father used to make similar calls on old friends, at farmsteads and at the large houses in the district, to ask the very same question." And although cider was now associated with the West Country, wrote Scott, Sussex had been renowned for the exceptional quality of its cider: the Bicknells were "still keeping alive one of Sussex's forgotten traditions."

Traditional cider making on the farm in post-war Sussex, at Hillgrove Common, near Northchapel, by Jim Bicknell, who had made his cider press by incorporating the central screw handed down through several generations. The photograph, taken by John Milner in 1947 or 8, was published in Hardiman Scott's Secret Sussex

Scott described in vivid detail how the Bicknells made cider from "all kinds and varieties" of apples collected in from the neighbourhood. Crushed in an aged cider mill, the apples came oozing out in a rich pulp into a tub, and were then placed in a press and covered with sacking, layer after layer, "dyed a ginger-golden colour with the juice of countless bushels of apples". About four bushels at a time went into the press, which Mr Bicknell had made himself in 1930. "But the great screw, into which the large metal handles fit to twist round and squeeze the apple pulp down into another large tub, has been in use for many generations," Scott noted. "The dull, cloudy liquid that flows into the tub is then already for putting down into six-gallon casks to mature." From the tubs, the apple juice went straight into casks and sealed down to mature, then delivered to customers immediately. "For the law only allows Mr Bicknell to sell direct from the press. As soon as it begins to ferment and has alcoholic content, then, if he wishes to sell it, he is subject to duty." In six to eight weeks, the cider would be ready for drinking, although Mr Bicknell recommended keeping it until the following spring. It was best drunk hot with a pinch of ground ginger, Mr Bicknell advised, and delicious it was, Scott agreed: "it was a drink for the gods – soft right enough, and sweet too, with the warm biting tang that the ginger gave it." Every bushel of apples made about two-and-a-half gallons of cider, and that year he made 640 gallons – although in a recent record year, he and his son had made 2,000 gallons.

Icklesham entrepreneurs: the Merrick brothers in 1961

All the while, more commercially focused apple enterprises were under way. To win, and keep, the attention of the consumers and wholesalers, growers developed marketing strategies, and Dick and Jack Merricks, growers at Icklesham, were pioneers. Just before the war, the family firm had introduced logos on its apple boxes – using an image of Hogg Hill Mill, the post mill on what was later to be Dick's land, in imitation of the display packs of imported apples. Colour coordination was also introduced on their stationery ("Dick Merricks, farmer, apples, hops and sheep"), on their boxes (marked "Dick Merricks – Sussex apples"), and even on their delivery lorries.[19] They went on to perfect the bulk storage of apples.[20] Bramleys put into store in October 1960 emerged in April 1961 with a loss rate of only 1 per cent. The cold storage chambers at Manor Farm packing house enabled fruit to be kept for six months. The apples were stored in easily handled bins, and moved using a Lansing lifter. The idea was to save on space, labour costs and packing costs. Producers were moving away from the old careful selection of good keeping apples, kept in cold, dry places, to a reliance on refrigerated storage buildings. According to the Merricks: "Using this method the work involved becomes almost congenial." [21]

Postwar, cider-making in Sussex received a boost as Jack Ward and Ian Howie set up the Merrydown wine company with John Kellond-Knight at Horam Manor. At 18 per cent proof, Merrydown vintage cider rapidly became very popular locally and in London: 5,000 gallons were pressed by the end of 1947, becoming 400,000 by 1955. Huge new oak vats, a bottle-cleaning machine

and a fleet of lorries and drivers were soon in place. The firm continued to use exclusively English fruit, including local suppliers from Sussex, and diversified into wine vinegar, apple-based compost, mead and a vineyard. Exports abroad also grew, as cider was re-branded as apple wine. In 1984 the firm's raw material – some 15,000 tonnes of apples – came almost entirely from Sussex and Kent. In the 1990s, as cider sales dwindled, Merrydown joined the alco-pop craze. By 2004, its non-alcoholic sparkling drink, Shloer, accounted for more than 70 per cent of turnover. But in 2005 Merrydown plc was purchased by the larger SHS Group and the plant at Horam closed, as manufacturing shifted to the west country.[22]

David Atkins was a prominent figure among apple growers in Sussex in the latter half of the twentieth century. He began his orchard enterprise near Pulborough in 1954, and vividly describes his experiences over 35 years in *The Cuckoo in June*.[23] He recounts with humour the hazards of apple-growing, from hailstorms and drought to wayward pickers. Clearly the stresses occasionally tested his temper to the limit, and one picker who was on the receiving end of his anger responded in verse, as Atkins remembers:

> Several nights of frost left one's nerves on edge. One of our best girls remarked on the bad frost damage and asked if I'd seen it.
> "Of course I bloody have," I snarled at her.
> Next day she handed me an envelope. In it was a short poem:

Picker poet: Sussex dialect used in spirited response to orchard owner: cartoon by Roger Smith, from The Cuckoo in June, 1992.

Apples measure up: a painting by Juliet Parnett, commissioned by David Atkins, shows workers at the grading machine still in use at Tullens Fruit Farm and now owned by Ivor Kiverstein, who has entertained many visiting schoolchildren by demonstrating the machine in action

David Atkins also recounts his efforts to combat pests and diseases in his orchard, illustrating that by the early postwar years, the use of intensive spraying had become the norm to combat insects and fungal diseases. Health and safety was not then an issue, and there was little protection for those doing the spraying, as at Dell Farm, Walberton, where highly toxic mercury was used, dissolved in water, between March and May each year. A hat, gloves or sou'wester might be worn!

A brief history of the orchard pest wars

Today, commercial, intensively-managed orchards are more heavily treated with pesticides – insecticides and fungicides – than arable crops, and are second only to vineyards in the amount of pesticides they receive, hectare for hectare.[24] Indeed, throughout the twentieth century, orchards have been at the forefront of pesticide R&D, and have become prime testing grounds for an ever-changing array of anti-pest preparations. Intensive use of pesticides in orchards began in the late nineteenth and early twentieth century, with the advent of huge North American fruit farms and the birth of apples as a global commodity. Before the 1880s and 90s, apples in the US and in Britain were mostly a minor crop on diversified farms, where the occasional failure of the fruit crop did not spell financial disaster. Apples with minor imperfections could be sold in local markets, and, by and large, pests had little economic significance.[25]

But as a specialised, large-scale apple industry began to develop first in New York State and then in Washington State, mass markets were sought. Productivity had to be high and labour costs low, favouring mechanisation. The apples themselves had to be flawless to withstand long journeys across America and beyond, to compete with indigenous, locally-grown apples in Britain.

The huge commercial monoculture plantations of Delicious in Washington State on the west coast of the US proved to be perfect breeding grounds for pests, inspiring a new breed of agricultural entomologists from the US Department of Agriculture, from the 1890s onwards, intent on finding chemical means to destroy the insects and fungi that threatened to blemish the nation's apple commodities. A host of new compounds were sprayed, over and over again, throughout the orchards, and if British commercial fruit-growers were to remain in business they had to follow this lead.[26] On bare trees, sprays of caustic soda and emulsified oil – various sorts of kerosene or paraffin – were designed to "burn, dissolve, suffocate or bedraggle any sucking insects there might be" in winter. But this radical treatment unleashed new problem pests. "The lichen and mosses vanished, the Canker and Woolly Aphis retreated, and the Apple Scab, Codling Moth, and, later, Red Spider came to the fore," wrote British plant scientist Ernest Large (1902-76) in 1940.[27]

To tackle leaf-eating caterpillars, trees were sprayed in the spring with arsenic compounds – variously Paris Green, tinted green by copper, or white arsenic or London Purple – a waste product of the synthetic dye industry. Spraying machines were refined, with better pressure control and nozzle design, producing a spray so fine that it would hang in the air like a mist and cover the foliage both evenly and economically. So equipped, American orchardists were ready for the attack on the Apple Scab fungus. Copper sprays became the favoured solution, with the so-called Bordeaux mixture, sprayed routinely, four times a year.

The American orchardists saw these developments in terms of reductions in the cost of production, Large argued, yet he presciently added: "but everything that the fruit-growers could gain by years of spraying could be lost in an hour by a slump of the market, a change of fiscal policy or a financial manoeuvre on Wall Street." The real thrill, at least for the scientific men, he said, "was an increase in the power of man to suppress the jungle and to defy the elements. Ultimately it was no less than that, for in the past the amount of Scab on the apples had depended upon the weather." In humid or wet weather, the fungus had spread, but was no problem in dry spells. Now, weather was immaterial: "one of the elements of chance in the cultivation of apples had been eliminated."

With all this backstory, orchards were primed to became vital testbeds for the burgeoning pesticide industry in the wake of the Second World War.[28] Apples were among the first crops to be treated with the new generation of powerful post-war agri-chemicals, the organochlorine pesticides such as DDT (dichlorodiphenyltrichloroethane).

Having triumphed over the enemy abroad, American agriculturalists set out on an eradication campaign at home. Armed with new and highly effective chemicals – such as DDT, unleashed against codling moth – apple producers at first imagined that the war against pests would soon be won. But DDT turned out to be such a good insecticide that it wiped out all the natural predators of pests, inducing outbreaks of European red spider mite and Pacific mite, organisms that had never been a particular problem before.[29] So growers ended up having to use chemicals in combination and sequentially, as new outbreaks occurred, or as pests developed resistance to the chemicals, with apples being sprayed between 12 and 20 times a year. Meanwhile, the quest for a "technical fix" for pest problems did nothing to address the ongoing and fundamental social and economic pressures on producers.[30]

British farmers and horticulturists quickly adopted these new agri-chemicals, especially fruit growers. "Orchards in Britain tend to be more heavily sprayed with pesticides, both insecticides (against codling moth, winter moth, capsids, woolly aphids to name but a few pests) and fungicides (against scab and mildew), than any other area in the country," wrote Kenneth Mellanby, a leading ecologist of the Nature Conservancy in 1967.[31] "Some fruit farmers are known to spray as many as twenty times in a season." DDT and a host of related compounds such as aldrin, dieldrin and heptachlor were soon widely used in English orchards.

In Sussex, John Nash of Costrongs Fruit Farm in Kirdford remember the days in the 1950s when the "orchards were grey" – with the intensively sprayed trees wiped clean of any trace of lichen or moss, which harmlessly grow on apple trees in more eco-friendly circumstances.

The American environmentalist Rachel Carson's landmark text, *Silent Spring*, published in the US in 1962 and in Britain the following year, exposed the unintended environmental consequences of this zero-tolerance approach to pest control. She explained: "I criticize the modern chemical method not because it controls harmful insects but because it controls them badly and inefficiently and creates many dangerous side effects on doing so."[32]

The organochlorine pesticides turned out to persist in the worms and insects in the soil in British orchards and fields, building up in the fat deposits of birds and small mammals, unleashing a toxic overdose that killed top predators such as kestrels, peregrines, sparrowhawks and foxes. It was the massive die-off of foxes in England in the early 1960s that first set alarm-bells ringing in the British establishment, as foxhunting enthusiasts in the House of Lords began to ask questions about these pesticides.[33] By the 1980s, Sussex farmer Michael Shrubb had conclusively linked DDT's extensive use in orchards to a collapse of sparrowhawk populations in Kent and Sussex.[34] By the mid-1980s, the use of DDT was banned in the UK.

Pest wars: Sid Marsh spraying from a tractor lift at Dell Farm, Walberton, in 1961, when DDT was still a widely used pesticide throughout Britain. Its use in orchards contributed to devastating declines in many populations of wild birds of prey

Organochlorine pesticides have been largely replaced by a host of equally powerful though less persistant organophosphates, as part of the complex mix of fungicides, insecticides and herbicides employed by many fruit growers. (See chapters 8, 9 and 11 for more discussion of pest control in modern commercial orchards.)

"Modernised" orchards change shape

Regular mechanised spraying with pesticides, picking and pruning were all made far easier, as the intensification of apple growing continued, by a profound shift in the size and spacing of apple trees in commercial orchards. In the 1930s, growers had begun to replace their old, full-sized "standard" apple trees that produced fruit over many decades in a non-intensive system with substantially smaller trees on semi-vigorous rootstocks. And then, in the 1970s, an even more radical transformation began, as growers began to move over to dwarfing rootstocks such as M9 which produce even smaller, bush-like trees. The most intensive of modern orchards may now grow more than a thousand mini-trees per acre rather than the 50 or so tall trees typical of a standard orchard. The dwarf "bush-trees" begin to bear fruit much more quickly than full-sized trees, but are replaced after just 10 to 15 years as maximum productivity declines.

This marked change in the nature of commercial orchards explains why it has proved difficult to compare the productivity of Sussex's orchards over the nineteenth and twentieth centuries. At the outset of our research, we wanted to find out when was the high point for orchards in the county – in other words, when did we produce the most apples? It turns out that what appeared to be a simple question is not so straightforward after all, because the nature of orchards has changed so dramatically over time. The

Hand-picked: Bob Simmonds harvesting apples at Walberton in the 1960s

newer type of orchard produces more fruit in less space, so simply knowing the size of an orchard is not a consistent guide to its output of fruit. What's more, some Sussex growers were quick to intensify their apple production, while others adopted a more conservative approach, such as the Ramuses, who began the shift to dwarfing rootstocks only after the hurricane of 1987 destroyed many of their big, old trees (see chapters 8 and 9).

By 1971 orchard producers in England were being offered grants to grub out orchards, in an attempt to increase the incomes of those remaining in the business, and more than 1,200 hectares were grubbed out by the end of that year.[35] Another massive sea change affecting apple growing occurred in 1973 when Britain became a full member of the European Common Market (later the European Union). By this time much of the commercial industry relied on sales to supermarkets, which had gained enormous influence in the 1960s, and the French Golden Delicious, with good keeping qualities, became their fruit of choice. It was a heavy yielding variety, relatively easy to grow and offered a uniform size, allowing easy and attractive packaging. English growers responded with campaigns to promote Cox's Orange Pippin and the Bramley Seedling.

Tom Mursell, from the Kirdford area, led a group of apple growers who took their protest to Brussels to the European Economic Community (EEC):

> We seemed to become the dumping ground for everybody's fruit and it decimated our market. If you look at the Kirdford zone, and we are under the clay influence, bigger fruits were coming in and we were hampered by our growing conditions. I would claim that was our downfall. We squawked about the EEC. I led a march in the Seventies in Brussels and was the cheerleader. "Who do we hate? The French. Why? Golden Delicious".[36]

Another impact of the EEC Common Agricultural Policy was the over-production which followed from the protected agricultural prices. These policies resulted in the "butter and wheat mountains" and "milk lakes". Quotas, introduced to stem this problem, meant that any apple production above the quota went into intervention and thence to cattle feed or fertiliser.[37] By 1996 farmers were again being offered grants to grub up their orchards under EU regulations designed to reduce the overall volume of dessert

The distribution of orchards in Sussex 1950

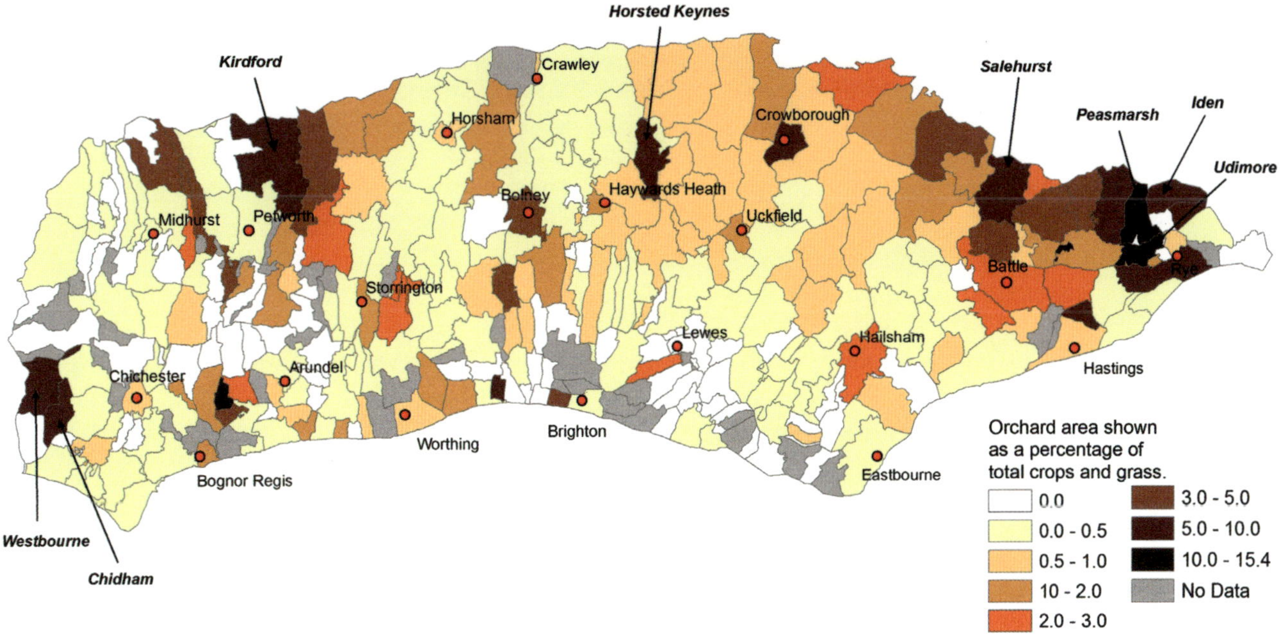

The distribution of orchards in Sussex 1980

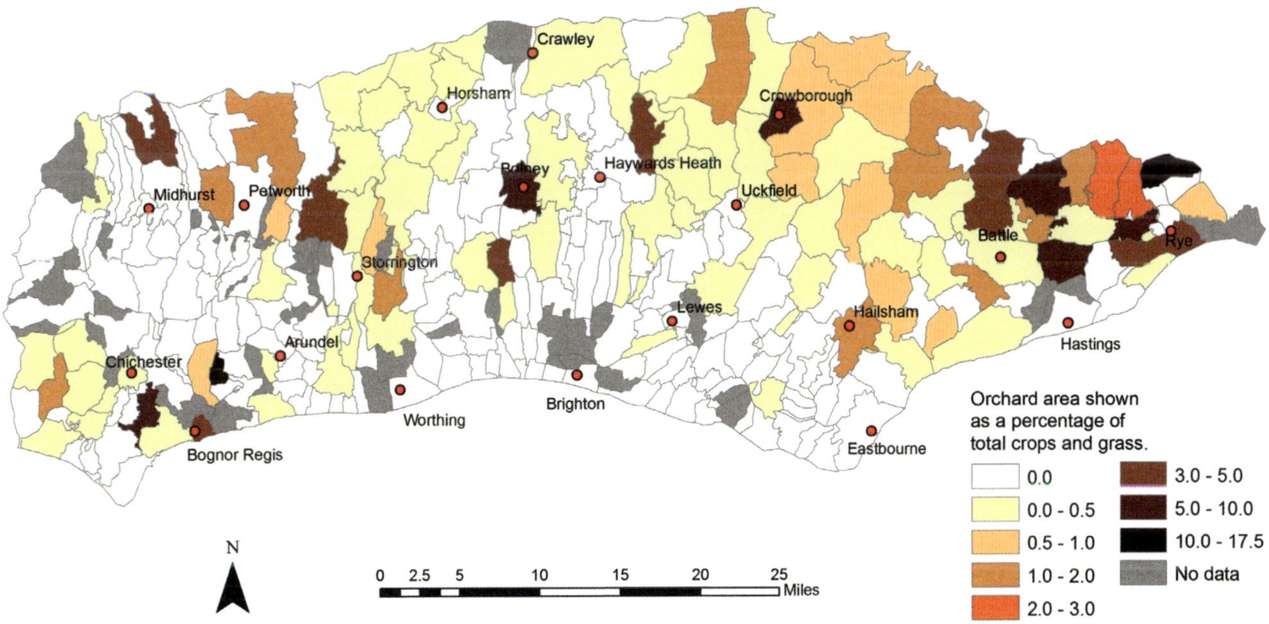

and culinary apples in Europe. Tom Mursell continued: "Then they encouraged us to grub up. If you went out of apple production entirely you could get compensation."

Across England, the vast majority of orchards were indeed grubbed up between 1961 and 1995 in response to agricultural subsidies that could be claimed for alternative, more intensive, land use and strong European competition compounded by imports from North America and the Commonwealth. In 1970, government statistics record 62,200 hectares (ha) of orchards in the UK (official data was now in metric units). Ten years later, this had fallen to 46,000 ha, and by 1997, only 22,400 hectares remained. And the decline continued – between 1997 and 2002, apple production in the UK as a whole fell further, by some 27 per cent.

By the early years of the twenty-first century Britain was importing 75% of its apples, even at traditional harvesting time. Even many cider apples were now imported cheaply from France.

In Sussex too, from the highpoint of 1955, orchard area declined steadily over the 50 years to 2005 – from nearly 2,900 hectares to 567 hectares – as farmers grubbed out their trees, and turned to more profitable crops or sold the land for housing.

Two maps, showing the area put down to orchards across Sussex, parish by parish, vividly show the dramatic loss of orchards between 1950 and 1980.

In 1950, at almost the peak of orchard activity in Sussex, orchards were commonplace. There were fruit orchards in nearly every parish in the county. They were particularly prevalent along the Kent border, in the area north of Petworth and Midhurst, in the far west of the county at Chidham, at Horsted Keynes and around Hailsham.

In 1980, orchards were in steep decline, and the situation continued to worsen into the next decades [38]. Amongst many others, Rory Clarke's orchard in Horsted Keynes was grubbed up around 1987, Dounhurst Farm at Kirdford closed in 1991 and so did Costrongs, one of the last orchards in the Kirdford area, in 2001.

The Kirdford growers' cooperative, finally disbanded in the 1990s, was a notable Sussex casualty of the EU quotas and government grants. In 2008 new houses were built on the land upon which the packing house and stores had stood. Street names such as Orchard Lane and Bramley Close are all that remain of an historic orchard-rich landscape. As a tribute to that once-vital rural industry, eight apple trees of old Sussex varieties have recently been planted, through the Local Fruit Futures project, on the village green, next to the community-run Kirdford Village stores which once housed part of the growers' cooperative.[39]

Grounds for cautious optimism

Today there are officially 715 hectares of orchard in Sussex. That's considerably less than the county's mid-century total of more than 2969 hectares. However this figure provides grounds for cautious optimism as, since 2007, government statistics show a small but steady rise in the area of land in Sussex covered by orchards – a rise of only a few per cent, but the first upturn in more than fifty years.

Orchard area in Sussex 1925-2010 in hectares

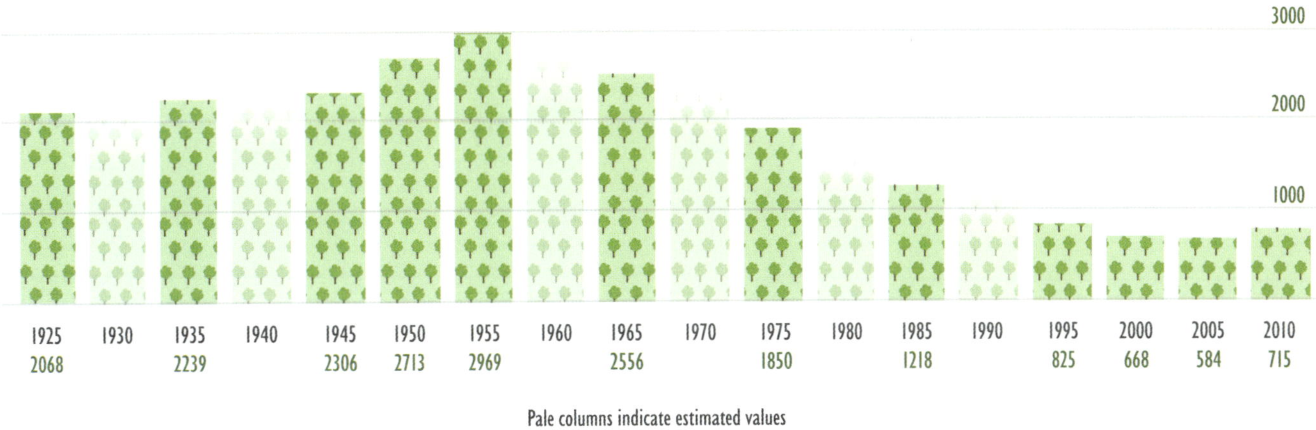

Source: Ministry of Agriculture 4 June Returns; and http://www.defra.gov.uk/statistics/foodfarm/landuselivestock/junesurvey

Growing apples in Sussex is still very much a going concern. In the next two chapters, people with many years' experience of orchard life tell their stories, and offer a wealth of perspectives on the past, present and future of orcharding in Sussex.

Endnotes for Chapter 7

1 V. Sackville-West, *Orchard and vineyard* (London: John Lane The Bodley Head 1921). From the poem 'April' (p. 51).

2 www. catalogue.communitysites.co.uk/assets/letterintheattic/.../LIA_72_5.doc. [accessed 22 November 2010]; Morgan and Richards, 277.

3 Ruby Dunn, *Moulsecoomb days: Learning and Teaching on a Brighton Council Estate 1922-1947* Brighton: QueenSpark Books 1990. http://www.queensparkbooks.org.uk/books/content [accessed 22 November 2010].

4 Janet Waymark, 'Streeter, Frederick (1877/9–1975)', *Oxford Dictionary of National Biography*, online edn, Oxford University Press, May 2009 [http://www.oxforddnb.com/view/article/95219, accessed 20 Oct 2010].

5 R. Kipling (Ed. T. Pinney), *The letters of Rudyard Kipling*, Volume 6 (1931-1936), University of Iowa Press, 2004, 210.

6 Mr Walter Guinness, Minister of Agriculture, HC Deb 08 November 1928 vol 222 .

7 J. Boyd Orr, *Food, Health and Income* London: Macmillan, 1936.

8 EWH Briault, *The land of Britain*: Vol VIII South-Eastern England: Sussex (East and West) London: Geographical Publications Ltd 1942, 519-51

9 See chapter 8, Peter May's interview with Tom Mursell in 2010.

10 John Nash talking to the Editor, 'A Wealden fruit farm' *The Petworth Society Magazine* 67 1992, 17-23.

11 *Midhurst and Petworth Observer* 1 February 2010.

12 EWH Briault, *The land of Britain*, 519-51

13 K. Murray, *Agriculture* (History of the Second World War Civil series) London: HMSO 1955, 381.

14 TNA, MAF 32/1010/103. National Farm Survey 1941-43 documents for Kingston-near-Lewes. Major Everett is noted again in the oral histories (chapter 8), in connection with Castel Mer farm.

15 G. Clarke, *The Women's Land Army: a portrait* (Bristol: Sansom and Co. 2008), 119-20.

16 J. S. Collis, *The worm forgives the plough* London: Vintage Books 2009 (orig, edtn 1973), 15 and infra. Originally published as *While following the plough* by Cape 1946.

17 British Library Maps 137 b.10.

18 Hardiman Scott, *Secret Sussex*, London: Batchworth Press, 1949.

19 R. Carr, 'Farmer's house style', *VADS* Feb. 1968, 38-40.

20 There is a memorial to Jack Merricks, who died in 1970, in Icklesham Church. Dick Merricks was the owner of Hogg Hill, the inspiration for the logo. The house, now restored since the early 1980s, belongs to Sir Paul McCartney and is used as a recording studio.

21 MERL P DX316 PH2/G4. Pencil notes on reverse of a photograph of the cold store.

22 Joan Thirsk, *Alternative agriculture*, Oxford University Press, 1997, 174, 241-2; *Sussex Country Weekly* 17 August 1985, 26; G. Wright, *Merrydown: Forty vintage years* Horam: Merrydown Wine Ltd, 1988. There are many photographs of the deserted site online, as for example at http://www.28dayslater.co.uk/forums/showthread.php?t=59894 [accessed 28 June 2011].

23 D. Atkins, *A cuckoo in June: tales of a Sussex orchard*, Dovecot Press, 1992.

24 Pesticide Usage Survey Reports, orchards and fruit stores in Britain, 2009, www.csl.gov.uk, now part of Fera, Food and Environment Research Agency; Pesticide Action Network Europe.

25 John Perkins, *Insects, Experts and the Insecticide Crisis: the quest for new pest management strategies*, NY: Plenum Press, 1982, p 19.

26 J.F.M. Clark, *Bugs and the Victorians*, New Haven, Conn: Yale University Press, 2009.

27 E.C. Large, *The Advance of the Fungi*, London: Jonathan Cape, 1940.

28 Perkins, ibid., p 21.

29 Mark L Winston, *Nature Wars: people vs. pests*, New Haven, Conn: Harvard University Press, 1997, pp 80-91.

30 Edmund Russell, *War and Nature: fighting humans and insects and chemicals from World War I to Silent Spring*, Cambridge: Cambridge University Press, 2001

31 Kenneth Mellanby, *Pesticides and Pollution*, London: Collins, 1967, p129.

32 Carson quoted in *Bugs in the System: redesigning the pesticide industry for sustainable agriculture*, ed. W. Vorley and D. Keeney, London: Earthscan, 1998, p 37

33 Norman Moore, *The Bird of Time: the science and politics of nature conservation*, Cambridge: Cambridge University Press, 1987, p 159..

34 Michael Shrubb, *Birds, Scythes and Combines: a history of birds and agricultural change*, Cambridge: Cambridge University Press, 2003, p 181.

35 House of Commons Debate, 7 December 1971 cc1109-10.

36 Tom Mursell, see interview in chapter 8.

37 When the free market price of specified commodities falls some way below the target price, the CAP intervention board intervenes by purchasing and storing the product. The commodity is then subsequently resold within or outside the EC, disposed of as food aid, or destroyed.

38 The table below allows comparison between West and East Sussex and with the much larger scale apple producer, Kent – the decline in production was dramatic in all between 1990 and 2005.

Areas of orchards in Kent and Sussex 1990-2005

	1990		2000		2005	
	hectares	growers	hectares	growers	hectares	growers
Kent	14,480	897	9,963	745	8,257	615
East Sussex	729	101	488	117	443	134
West Sussex	388	91	180	86	141	83

Source: Ministry of Agriculture 4 June Returns

39 The shop, opened in May 2010 and launched by actor Julie Walters, won the *Daily Telegraph* award for the best village/corner shop in Britain in 2010.

8. Orchard Stories

8

Orchard stories

The changing nature of orchards in Sussex over the last 80 years comes alive in the following pages through the words of growers and others whose stories convey something of the sheer variety and richness of the way orchard fruits have impacted on our cultural and economic, as well as physical, landscape.

The following transcripts are edited versions of interviews conducted by Peter May during 2010 and 2011. For the interview with Charles and Nicholas Ramus he was joined by Anne-Marie Bur, who also talked with Gladys Marshall of Kirdford. Some of the interviews were recorded while in others (Peter Sharp, John Nash, Fiona Durling, Matthew Wilson and Eric Rowlands), Peter made notes from which the interviews have been reconstructed, and all texts have been approved by the interviewees.

The interviewees were:

- Rory Clarke at the Ashdown Land Company near Horsted Keynes
- Peter Sharp at Castelmer Fruit Farm, Kingston
- Charles and Nicholas Ramus, Barons Grange Farm, Iden
- John Nash of Costrongs Fruit Farm, Kirdford
- Gladys Marshall of Kirdford
- Tom Mursell at Dounhurst Farm, Kirdford
- Fiona Durling at The Dell Orchard and Nursery, Walberton
- Graham Love at Greenways Fruit Farm, Herstmonceux
- Tom Maynard at Maynards Fruit Farm, Ticehurst
- Matthew Wilson of Oakwood Farm, Robertsbridge
- Tony Eales at Tendrings Fruit Farm, Magham Down near Herstmonceux
- Eric and Janet Rowlands at Great Sanders Fruit Farm at Cripps Corner
- Jim Buckland at West Dean Gardens near Chichester

Rory Clarke at the Ashdown Land Company

The Ashdown Land Company was started at the time of the agricultural depression in the 1930s when 170 acres of apples, pears, plums were planted on the Broadhurst Estate near Horsted Keynes. There was also soft fruit which was made into jam and marketed under the Blue Smock label. St Martins, a large producer of marmalade, later took over this side of the business. As many as 20-30 different types of apple were grown on the estate together with plum orchards of Purple Pershores, Giant Prune, Victoria and Gages. The company used wholesalers to market the fruit which was sent to Covent Garden, Brentford, Brighton and Southampton.

My father started the firm in the 1930s at the time of the depression, a terrible time when farms were being handed in. He had been interested in fruit so started planting up the farm with advice from Long Ashton. He grew a lot of soft fruit together with apples, pears, plums and peaches. Then in the late 1930s he started another company which did canning for the soft fruit. This was based in the present day industrial estate where on the extreme left we had a packing line and stores for the fruit.

How many acres of fruit were there?
The fruit was about 170 acres. Here the soils are Ashdown Sands with a clay cap, though the orchards were planted on the best drained fields. The clay runs through here [the estate] and is not fit for anything except pasture. At the other end we have light sandy soil which are grades 3-4 [poor soils]. The estate is on three valleys and on the bottom of the valleys are hammer ponds.

During the war St Martins came on the scene who took over the jamming part of the business. St Martins were a company which made jam and marmalade. They gave up the buildings in the 1950s then there was a hatchery and poultry. We also put in more storage and went into bulkbins. During the 1950s there was about 20 acres of soft fruit which was a non-goer as it got in the way of the apples and we could not get enough labour to pick it. The original trees were large on M1 or M2 then when I came back we had MM106 and M9.

Rory Clarke took over his father's thriving orchard, which lasted more than 60 years

My father John Clarke probably had 20 to 30 different varieties of apple and it was up to me to change that to early, mid-season and late varieties. The earlies were George Cave, Discovery and Ashdown Seedling. Ashdown seedling he always said was produced from a bright red McIntosh apple that he was eating. He kept some of the pips which he then grew on in the kitchen garden at home. We thought it a nice early apple so we propagated it and sold in competition with Discovery. They are similar in taste and to look at as well. The tree cropped regularly and it was not a bad apple though a bit sour. We were still growing it up until the time we gave up in 1991.

George Hall, packhouse foreman, and pickers harvest Victoria plums in the early 50s

What sort of pest control was used?

In the old days pest control was tar oil wash in winter then we used lead arsenate, lime sulphur, DDT, BHC and Lindane which were all permissible in those days.

How many people were employed?

We had about 20 people. They were all housed on the place with a foreman, a manager and we had a packhouse foreman. They were pretty well employed all the year round, picking from August, then packing until February and March. At other times they worked on the farm. We also had a farm manager and a farm foreman. We had two dairies and we grew a bit of corn. The farm was about 700 acres.

What were the main markets for the fruit?

We mostly did Covent Garden, Brentford, Brighton, Southampton and later we did Dartford. We used wholesalers. There was a firm called Hales Dancy, Plupart, Braddenhams and Addenhams. We had our own lorries. The railway was used to bring fertilizer down and we had some old carriages where people stayed for picking. In the

1920s and 30s they used to come down from London and have these camps, but in my day that sort of thing had ended. The workforce was mainly female but when they wanted to work from home these ladies disappeared. We used to get foreign workers in and the Poles were brilliant. We also had students.

The orchard was cultivated until the 90s. What happened was that we had a gale here in 1987 which blew all the fruit off and the trees went down. The following year there was a glut and prices were low and we said to hell with this. We could not get enough labour so we decided to call it a day and get a grant to remove it ... very sad, I miss my fruit.

Peter Sharp at Castelmer Fruit Farm, Kingston

Peter Sharp was a conscientious objector during the Second World War and worked on the land at Ripley in Surrey. Later he accepted a job with Major Raymond Everett in Kingston near Lewes at a time when the village had three orchards growing a mixture of apples, soft fruit and vegetables. The apples at Castelmer were then mainly grown as densely planted cordons, with glasshouse crops of tomatoes and bedding plants. The farm closed in 1997.

After the First World War there were three orchards at Kingston which grew a mixture of apples, soft fruit and vegetables. Col Smith, a retired army officer, had made land available in Kingston for sale to ex-army officers to cultivate.

Major Snoxhall at Kingston Ridge Fruit Farm employed 20 men and sold direct to a Brighton shop, Ron Turner grew soft fruit, vegetables and a few apple trees. Then in 1922 Raymond Everett the owner of the Castelmer site started growing salad crops organically or as it was then known "compost grown".

Peter was a conscientious objector during the war and in 1941 was working on the land at Ripley in Surrey. He answered an advert for a "compost gardener" at a small progressive school in Forest Row. Compost gardening was an early expression of the organic movement and its supporters "used no artificial".

After two years he placed an advert in The Grower magazine as Grower Foreman and received five replies. Peter accepted a job with Major Raymond Everett in Kingston. Peter, as a conscientious objector, was surprised to be employed by an ex-army man. However, Everett was a member of the "reconciliation movement" which Peter thinks may have changed his opinion of pacifists.

Castelmer came up for sale two months after Peter started work and was bought by Guy Herriot who lived in the village. A colonial rubber planter who had been imprisoned by the Japanese, he never did any of the practical work, but owned a Jaguar car and regularly visited his club in London. Peter describes him as a Liberal anti-vivisectionist.

Three thousand densely planted cordon apple trees were planted in 1932 with varieties such as Cox, Beauty of Bath, James Grieve, Lord Lambourne, and Ellison's Orange. The apples were sold in Lewes, Brighton and also a wholefood shop in Baker Street, London.

Guy Heriot provided capital to plant further apple trees on a sloping site with poor thin soils, and £500 to buy a glasshouse in the hope that Peter "could make it pay." The trees on the hillside never grew well because the soil was too thin.

Guy Heriot had put himself into debt to purchase land and a bank agreed to support this debt provided Peter Sharp was made a partner. A legal agreement was put into place which made Peter a partner with Guy Heriot. Three-quarters of an acre of Castelmer was then sold off and half share of £2000 given to Peter. Then in 1966 the bank wanted its money back. To prevent the loss of the land a solicitor devised an agreement that the Shern Hall Methodist Building Society provide a mortgage of £25 a month for 25 years. Peter then became the full owner.

Peter was also interested in producing crops under glass. Tomatoes were grown, which were grafted onto disease-resistant rootstocks and these plants were grown for sale and to be planted out as a fruiting crop. Lettuce, cucumbers and bedding plants were also grown. Mushrooms were an important crop and grown on two raised benches. The method used was that the mycelium was grown on a layer of horse manure which overlay a chalk, soil, compost mixture. There were six glasshouses which were used until 1997 when they were sold for building land.

Facing page: Peter Sharp, one of thousands who worked the land rather than take up arms came into orchard work initially via an ex-army member of 'the reconciliation movement'

Charles and Nicholas Ramus, Barons Grange Farm, Iden

The 250 acres of top fruit at Barons Grange Farm is the largest commercial orchard in Sussex. Up until the 1970s the Ramus family planted standard trees of Bramley, Newton Wonder and Laxtons Superb. However, the storm of 1987 felled 50-70 per cent of the standard orchards and this brought about a change in orchard management. The orchards were intensively replanted, mainly with Bramley and Egremont Russet on dwarfing rootstocks. Then 10 years ago, a neighbouring farm was bought containing 80 acres of Cox. C&M Ramus and Sons supply the wholesale markets which in turn supply greengrocers, hotels, hospitals, farm shops and caterers. The firm also supplies apples for processing and juice for the Copella brand.

CR = Charles Ramus (son)
NR= Nicholas Ramus (father)
Interview by Peter May (PM) and Anne-Marie Bur (AMB)

Charles Ramus and Nicholas Ramus, son and father, run the largest commercial orchard in Sussex

PM: Could you give a brief introduction to how the orchards were set up?

NR: Well, the orchards were originally planted by my grandfather on my mother's side. The family name was Bates, C.E. Bates. They, with their brothers, owned land in pieces from Northiam to Romney Marsh. A pretty large concern. This particular farm, from my grandfather's point of view, was given to my mother, as a life interest to go onto her children. She got married in 1930 to my father who at the time was in the merchant navy.

PM. What sort of farm was it then ?

NR: Top fruit and hops. C.E. Bates planted the majority of the fruit in the 1920s. They were Bramley, Newton, Laxton and they were on a 40 x 40 ft plant. They planted plums. When the trees were really young they planted gooseberries in between and later took them out. Then took out the plums, then the Newtons because they could not sell them. There were Bramleys back from the 1920s which take a long time to establish.

PM: So all these trees were planted on Crab C [a vigorous rootstock] or the equivalent?

NR: Yes, they were all the traditional big Bramley trees but they took 20 years to establish and that is why they planted plums in between. And then when the Second World War came they ploughed and grew crops underneath the trees, because you can still see the ridges. They were levelling out the land between the trees one day and Pop said I do not mind if it is a bit wavy . . . and it is still a bit wavy to this day. After the war that all petered out except for the sheep and we traditionally ran sheep through the orchards through the summer. And that went on until the 1980s. The hurricane triggered the change and the next generation.

AMB: It wasn't market forces and the import of apples from Europe?

CR: We were planting standard trees until the 1970s and my father would not hear anything different. After the hurricane young bush type trees were put in to get production back in quickly. 75 per cent of the young standard orchards were gone overnight by the storm and 50 per cent of the established ones. Some cocky sod [referring to himself!] came back from college and told them how to do it. My mother passed away three weeks after . . . she was shocked by what she had seen . . . it destroyed her.

PM: So you had to grub up the stumps?

CR: Yes, I sat on a bulldozer all winter and had fires. We managed to get hold of some trees and get them in during the winter of 1988.

PM: These new plantings were on MM106?

CR: Yes, we had to get what trees we could. From then on we planted MM106 with M9 interstocks which is quite late to be transferring to that type of tree. Then 10 years ago a neighbouring farm came up for sale which was all Cox. These were intensively planted on wirework so we bought 80 acres of Cox and kept it running how it was but mechanically pruned it. There is a demand for Cox, it is one of the two varieties they look for dessert wise on the wholesale market, but we had to grow it cheaply and the only way we could do it was by cutting labour and mechanically prune them. Everyone said that you are absolutely insane. We have done it and made it work financially. Anyone who says that they grow Cox and sell it on the wholesale market and prune it by hand, they are losing money.

PM: So this a series of rotating blades?

CR: The French developed it for vineyards and then they realised you could do this

for fruit. 10 years ago they had got a working model with small enough blades and fine enough giving a good enough cut . . . 10 saw blades in all on a frame . . . then every four years we hand prune and thin them down.
NR: we are at a crossroads now.

CR: The trees are getting old and are well over the age that people would keep then agewise . . . they are 30 years old which for M9 is top end.

PM: So what variety will you grow?
CR: We have concentrated on the wholesale market and we want to grow apples which people look for as being English . . . Cox and Bramley. Pears, Conference, Comice, Concorde. But dessert-wise it is Russet and Cox.

AMB: You mentioned that this was a crossroads in the business. What decisions will you make?
CR: We will be looking at what the wholesale trade will want. On the dessert side most people grow Gala for Sainsburys or Tescos so there will be some in their packhouse for the wholesale market. The funny thing is we are thinking of growing Cox. It has a brand name, everyone else are thinking of getting out of it so maybe it is the variety to stay in and the other one is Red Windsor.

NR: On the traditional Bramleys you did not pick until the fourth week of September. Now because we can get the size we are picking in August.

AMB: Why are the Bramleys ripening early?
CR: The smaller trees ripen earlier and produce bigger fruit. People do not like any red on the Bramley, they like them green because people shop with their eyes.

PM: Could you discuss the coldstore that we looked at earlier?
CR: The new coldstore is a thousand tons which is just over 3000 bins in five rooms which are all controlled atmosphere store. So that they are refrigerated and you just put in the temperature you want them to run at. They are using fairly green technology . . . there is no such thing as green refrigeration but they use the minimum amount of power to produce that. There is controlled atmosphere where you have a scrubber which produces the atmosphere you want . . . We can store for 12 months Bramley and Jonagold if we want to, no problem, touch wood.

[Note: Refrigeration in coldstores, as described by Charles Ramus, combined with controlled atmosphere storage, dramatically extends the storage life of apple crops. The respiration of the apple is reduced to a minimum by reducing the air temperature, keeping oxygen and carbon dioxide levels low and removing ethylene, a gas that is naturally produced by fruits which encourages ripening, using "scrubbers" that absorb particular gases and remove them from the atmosphere.]

John Nash, production manager at Costrongs Fruit Farm, Kirdford

In 1925 Bill Fowle purchased Costrong Farm just outside Kirdford at a time when orchards were seen as the future of local production. He planted Costrongs with Cox and Bramley on vigorous rootstocks, with the Bramleys forming enormous trees needing ladders to pick the fruit. Costrongs became one of the founder members of the Kirdford Growers Co-operative, the first growers' co-operative in the country which at its peak had 20 members with extensive packhouses and coldstores in the village of Kirdford. The farm was later replanted with trees on dwarfing rootstocks and tile drained by its next owner Colonel A.H. Cave. However, on the heavy soils of the Kirdford area it was always difficult to grow fruit to a uniform size that the supermarkets demanded and production declined. The co-operative packhouse was closed in 2001 and Costrongs was the last orchard in the area to be grubbed out.

Sally Cave, owner of Costrongs Farm and John Nash, Manager, leading players in the Kirdford Fruit Growers' Co-operative

Colonel AH Cave bought the farm in the 1950s, replanting the orchards with apples on dwarfing rootstocks which enabled more intensive production and greater yields per acre. He tile drained the entire farm and built two reservoirs. As the farm was in a frost hollow he installed 60 acres of overhead frost protection. This was necessary in an area where one in five of the crops could be lost with late spring frosts. John Nash noted: "Kirdford went down to –10°C one night and the next morning the orchard looked spectacular with icicles covering the branches. Some damage was done with branches breaking under the weight of the ice." A million gallons a night might be sprayed onto the trees maintaining the temperature just above freezing and saving the crop.

Costrongs Fruit Farm – modern-style orchard, where many pickers found work during the harvest – below they are gathered for the annual pickers' picnic held on the final day of the harvest

Colonel Cave grew predominately Cox using the latest spindle or hedgerow planting systems. Huge crops of variously sized fruit were sent to Covent Garden, Bristol and Cardiff by rail from Horsham station, with British Rail providing returnable wooden boxes padded with straw. In common with other Sussex producers lead arsenic and sulphur were used as fungicides with lime washes applied during the winter. The removal of the moss, lichen and some of the aphid eggs "cleaned the tree up" and was considered good orchard practice.

Most growers in the co-operative made a good living and sold the entire crop before Christmas when southern hemisphere fruit was allowed onto the British market. On the heavy soils of the Kirdford area, growers found it difficult to grow Cox to the uniform size and quality that the supermarkets were increasingly demanding. There was also competition from orchards on the better soils of Kent and increasing imports of high quality fruit from Holland which got a good price. The smaller apples of Kirdford, though a good colour, fetched a lower price and were sometimes not worth picking if the costs of marketing, storing and packing were considered.

Colonel Cave died in 1983 leaving his widow Sally to continue the business with John Nash as production manager.

A wider range of apple varieties was added to the cropping in the Costrong orchards such as Egremont Russet, Discovery and Worcester Pearmain together with the pear varieties Conference, Comice and Beurre Hardy. Many additional varieties were also sold in the farm shop including Adams Pearmain and Orleans Reinette, as customers showed an interest in the older varieties. Glasshouses were used for the production of winter and summer lettuce, tomatoes, cucumbers and this was later expanded to include peppers, aubergines, grapes, strawberries and bedding plants.

During the 1980s "integrated crop management" was introduced, replacing the blanket spraying of insecticides and fungicides of previous years. The latest controlled atmosphere technology was introduced into the coldstores. Despite these changes, apple production in the Kirdford area declined. The import of high quality fruit from Europe and the southern hemisphere undermined the market for the Kirdford fruit, which tended to be too small. The heavy clay soils and the possibility of late spring frosts had never made the area ideal for apple production, and now the low cropping levels and increased competition from imports were combined with a decline in the availability of part-time labour. All these factors deterred investment and gradually the Kirdford orchards were grubbed out. The co-operative packhouse and coldstore was closed in 2001 and the Costrong's orchards were themselves grubbed out, the last of the Kirdford area.

Gladys Marshall of Kirdford, with John Nash and Sally Cave

Gladys Marshall first came to West Sussex as a Land Girl at the age of seventeen, one of the eighty thousand women drafted into farm work to replace the men who had enlisted in the armed forces. The work and the place suited her. In fact she married the farmer's son. They have been married sixty-three years.

Gladys Marshall began working on the neighbouring Costrongs Farm as an apple picker so long ago she is not quite sure of the year but while her children were young so, it seems, at some stage in the nineteen fifties. Pickers brought their young children along in those days. She picked apples every year after that until the nineties. For many years pickers climbed tall 30 rung ladders to get to the fruit. When new lightweight aluminium ladders were introduced there was always a bit of a race in the

Gladys Marshall (left) enjoyed 'ladder work' while other pickers preferred to keep their feet on the ground

morning to get one of these rather than the heavy wooden ladders, Gladys remembered.

Pickers worked in pairs and regular pickers co-operated. Some who were unable or afraid to mount ladders were allowed to pick the bases while others might follow up and do "the tops". Gladys said her daughter liked "ladder work" and would happily do the tops. Falls were infrequent but far from unknown. John Nash, who was Manager of Costrongs said "You weren't an apple picker till you'd fallen off your ladder!" and he and Gladys between them recounted the story of a rather plump Polish picker who was five steps up a ladder picking Bramleys when she leaned too far and "came off sideways". Momentarily immobilised she lay on the ground and asked for "Wa'er" – "We thought she meant water and brought some but she was calling for her husband Walter who was picking somewhere else . . . we couldn't help giggling . . . and then, as though that wasn't enough, a fellow up top picking let go of a Bramley and it caught her smack on the forehead... she wouldn't go up more than two steps after that."

Another regular picker was a widow bringing up five children on her own. She had a jacket with a lot of internal pockets which tended to fill out as the day wore on. After visits to her motorbike and sidecar she slimmed down again. One day at the end of picking the vehicle toppled over, and her rescuers found the sidecar so full of apples that it might well have contributed to the mishap. Remembering this incident was another occasion for lots of laughter. No mention was ever made to the picker of the apples, in the jacket or the sidecar. "Well, we knew her situation" was John's only comment.

Gladys Marshall left written evidence on her picker's bag of long service in the orchards of Sussex (which is how we found out about her...)

Pickers were paid by the hour, not by quantity (piece work) because it was important that all the fruit was picked and not just the easiest and quickest. Gladys said it was a matter of pride not to leave any fruit behind. John said Gladys was a good picker. "You could rely on her – a poor picker would leave the tops behind." Each picker had their own bag. Over the years apple pickers' bags have been made of all kinds of materials including wood, canvas, aluminium and plastic. "Wood was best" according to John and Gladys agreed. Hers hangs still in the shed. On it Gladys had written "Glad's bucket. Forty years hard labour and loved every minute of it."

The highlight of the pickers' season was the pickers' picnic on the final day of the harvest, provided by Sally Cave, wife of the owner. Gladys used to regularly contribute an apple pie. Wine was provided too. "After a few years we gave up sending them back to work in the afternoon. You could hear giggles all over the orchard and no work got done," Sally remembered, with no trace of regret.

Tom Mursell at Dounhurst Farm, Kirdford

Orchard empire: an old aerial photograph shows the extent of the Kirdford orchards in the 1950s. Today, the trees have all been grubbed up

In the 1970s Tom Mursell of Dounhurst Farm near Kirdford led a march at the EEC headquarters in Brussels demonstrating against the import of large quantities of European and southern hemisphere apples that were decimating the English apple industry. The demonstrators shouted out: "Who do we hate? The French. Why? Golden Delicious!" Although the English apple industry was galvanised into action, little could be done when other countries could grow larger quantities of higher quality fruit that the supermarkets required. The effect of opening up the English market to foreign imports led to the end of apple growing in the Kirdford area. Dounhurst Farm had been bought in 1932 by Tom Mursell's grandmother and planted with 120 acres of apples and pears. His father, Sir Peter Mursell,, returned after the Second World War to work on the farm, and joined the Kirdford Growers Co-operative.

My grandmother was widowed and lived up in Cheshire and in 1932 decided to move down this way and bought 230 acres of Dounhurst Farm at £40 per acre. She promptly planted 120 acres of apples and pears. She was a stickler for detail and when trees were planted she would go around and tug the trees and if they were loose she would ask them to do it again. My father, Peter Mursell, had gone to Cambridge and got a double first in agriculture. Meanwhile Kirdford Growers had formed and Dounhurst was in the second tranche. Most had came back from being in the colonies or tea plantations and through Dr Hatton at East Malling realised that they could grow good quality flavoured fruit in the Kirdford area. Flavour was the key to growing Cox in this area. They used to meet weekly in the pub and have a chat. They soon realised that they each saw the same salesmen every week. With an amazing leap of imagination they thought they would form a co-operative and buy one of everything instead of four or five. Land was purchased in Kirdford and the packhouse created. They took it in turns monthly to be managing director and it was the first apple and pear co-operative in the country.

Tom Mursell began selling cans of his own apple juice in the 1980s

Did you share machinery or was it mainly a marketing co-operative?

No it was mainly marketing and storage but they also had a technical advisor whom they employed between them. They did join together to buy spray equipment to get a better price, but the early spraying was done from fixed standpipes. There were fixed underground spray mains and every 80ft there would be a standpipe and each chap would have a 60ft hose with hand lance. They'd sometimes put a sou'wester on but would not worry too much. You would plug the hose on then turn tap on. The central mixing pot at the pump house was where Bert Warner our foreman would be. He would chuck bags of fungicide into the vat which would be pumped under pressure around farm. All spraying in those days would be manually done. We were using lime sulphur and also nicotine as a drench. When they pumped nicotine the sprayers would come back yellow. There was no worry about health and safety.

Were the standpipes used for spraying also used for frost protection?

No. In the 1940s frost protection was done with smudge pots. They held about 6 gallons of diesel with a chimney on it. They used about 400 to the acre. The alarm [low temperatures] would go off and father would rally the troops. They would have paraffin with a wick and go and light the pot. There would be a black cloud over the orchard . . . the pollution! We had a 12,000 gallon tank of diesel which would be filled sometimes once a week if we had a bad frost period . . . but it was only 2d a gallon. My father tried an aero engine with a propeller, with the chaps driving around the farm trying to stir the air up. He also had a jet engine but that did not make any difference. He found that the hot air just went straight up.

In the late 1950s we put in two large reservoirs of eight acres holding 20 million gallons and we put in 60 acres of irrigated frost protection. At Dounhurst we were our own catchment area so that we sprayed overnight and it came back into the reservoir.

What was the effect of the European Community?

It opened up apple production as a world commodity and we seemed to become the dumping ground for everybody's fruit and it decimated our market. If you look at the Kirdford zone, and we are under the clay influence, bigger fruits were coming in and we were hampered by our growing conditions. I would claim that was our downfall. We squawked about the EEC. I led a march in the 1970s in Brussels and was the cheerleader. "Who do we hate? The French. Why? Golden Delicious!" Dan Neutboom, head of the NFU fruit committee galvanised us into action. At the time we were terribly British . . . "Not much we can do about the EEC".

Could Kirdford Growers do anything about this?

No, not really. The more productive and efficient orchards in England found a niche which was growing Cox. From a marketing point of view you could define English production with Cox and Bramley. They were our uniques. Other countries could grow larger quantities and higher quality fruit because they had more sunshine, while we were at the northern end of it. Some growers went the high input, high output route and some succeeded, though they went through their pains at a time when the industry was crippled (by high imports). England was left with just a few pockets of orchards.

Fiona Durling at The Dell Orchard and Nursery, Walberton.

Until the 1980s The Dell Orchard and Nursery produced a wide range of cut flowers, vegetables and apples for farm sales and the Covent Garden market. On this mixed holding there were also glasshouses for tomatoes and early chrysanthemums with peaches trained along the interior walls. During the 1950s 10 men were employed on a full- time basis and 10 women part-time with extensive use of tractors and cultivators.

Fiona Durling as a child (left) with her mother and brother

By 1959 there was a wide range of production at The Dell Orchard Nursery with cut flowers of all types from Gladioli and Chrysanthemums with Sweet Peas to Helichrysum, Zinnias, Statice, Golden Rod, Montbretia and Achillea. Glasshouses grew tomatoes and peaches were trained along the walls. Along the south coast peach growing was a local speciality and the grandfather of Fiona Durling had grown them under glass at Shoreham-by-Sea. Because there were no imports of peaches during the Second World War they were sold at £1 each and sent up to Epsom for Derby Day. There were fields of Brussels sprouts, cauliflowers and onions with Leveller gooseberries and an orchard on the western side of the holding. A sales leaflet describes the 17 different "quality picked apples" with early varieties such as Discovery and later season varieties such as Charles Ross and Egremont Russet. There were also three varieties of cooking apples with Grenadier, Bramley and Edward VII.

Fiona Durling displays the meticulous planting records in the highly productive orchard and nursery at Walberton

Produce was either sold from the nursery or loaded onto early trains for London at Barnham station. There was rivalry between the nurseries to have the best possible presentation. Reg Lee who worked for the wholesaler Poupart's at Long Acre remembers how the Barnham and Brighton trains would arrive in London and have to be unloaded and the produce got ready in the market for the next morning's trading before they could go home.

John Mills, the father of Fiona Durling, vividly writes about the effect of the war while working at the Dell Nursery:

> "Spat, spat all around us – we look up to see several Junkers 87 winging away from Tangmere airfield and heard the thunder of bombs exploding on the runway. All of us on the market garden that Friday in August, 1940, started to run to the nearest ditch to jump into, away from the machine gunners in the German planes. We found the old boys from the 1914-18 trenches already there, and they soon showed us how to keep our heads down to save ourselves. We looked up to see a squadron of Hurricanes wheeling in to attack the bombers, shooting down several over Selsey and Witterings. That was the start of the real war in West Sussex, with a great deal of air activity and gunfire over the next few months. We continued to work, being in or out of ditches, to grow

> produce of fruit, veg and salads, on the land, being told on the radio that our efforts would save a seaman's life because ships bringing food were sunk by U boats."

Even after the war, when rationing was extended, every inch of the nursery was used for food production, though cut flowers continued to be grown under the apple trees to relieve the drabness of postwar life. Wool shoddy – a by product of the wool textile industry – was used to improve the land, and Fiona Durling remembers how it turned the nursery blue when it was spread on the ground. Guano, imported from South America, was also used as a nitrogen-rich fertilizer as well as farmyard manure and hoof and horn.

At its greatest extent the nursery and orchards covered 30 acres with three-quarters of an acre of glasshouses and coldframes. A coldstore was added in 1960. Machinery was widely used with a John Brown tractor, rotary cultivators and a Colwood Motor Hoe.

In 1964, however, a hailstorm destroyed the glasshouse with a loss of many of the bulbs planted there for cut flowers. To pay for the repairs to the glasshouse half of the nursery land had to be sold. Fiona Durling relates that the nursery continued until the 1980s by which time it was only employing her father, his wife plus an aunty and a "youth".

Graham Love at Greenways Fruit Farm, Herstmonceux

Greenways Fruit Farm stands on a Greensand ridge in the Low Weald a few miles from Herstmonceux. Graham Love has been growing fruit here since the mid 1970s when he worked with his father producing apples and pears. The original orchards were planted in the 1950s by his father, with apples grown as large bushes on a vigorous rootstock and an area of pears which took 10 years before they cropped well and then formed one of the best pear orchards around. Graham had done a three-year course in Horticulture at Hadlow College and then did voluntary service in Africa before returning to Greenways. The orchards grow Gala, Braeburn, Elstar and Jonagold for the supermarkets and also a wider range for the retail trade and farmers markets. There is also a small area of cherries which crop prolifically.

My father came here in 1950 in his late 20s. His parents were mixed farmers in Kent and they wanted him to continue mixed farming, but he suffered badly from hay fever. It was cheaper in Sussex and he hunted for quite a long time to find a suitable site. Here there is a Greensand Ridge which is a very hungry soil based on the Tunbridge Wells sands and needs a lot of feeding but it is very good for horticulture. It was farmed by a tenant farmer through the war and was very run down.

Graham Love runs Greenways Fruit Farm with a team of workers

Was it a livestock farm?

Yes, the owner of the land wanted to sell the land and throw the tenant off. My father took it over. His parents paid for the land and he had a legacy from his uncle which paid for the trees and machinery. He relied on the bank for the first six or seven years. In those days the trees were planted extensively in order to cultivate and mow in both directions.

How do you mean extensively?

Wide apart. Yes, the trees were planted 17-foot square so tractors could go all the way around them. Hedgerow planting only came in during the 60s when there were herbicides.

Your father planted apples mainly on M2 [a vigorous rootstock]?

Yes, apples on M2 and pears on Quince A. The pears took ages to crop. They were planted in 1950 and did not crop until the 1960s. There is a saying that you plant pears for your heirs and this orchard did not crop well until the 1970s, 20 years after planting. It eventually formed one of the best pear orchards around and we won trophies and prizes. It is still producing now.

In the early years he was marketing everything through East Sussex Growers, a co-op at Polegate. The trouble was they would take everything, so growers sold the best grades themselves and sent everything else to E. Sussex, so the co-op sunk. My father then sent the crop to East Kent Packers who were more disciplined and demanded everything, including the best. As the crops came in, he developed a system of bulk bins which speeded up orchard harvesting by cutting out unnecessary handling and provided short-term storage. He had to pay to join East Kent Packers to pay for our share of packing and coldstore facilities so we never developed our own packing and coldstore facilities.

Team work: Graham Love is second from right

You worked with your father as a partner?

Yes, I left school in the mid-60s, worked on various horticultural establishments in England and Holland, did a three-year sandwich course at Hadlow College then went and did voluntary service in Africa co-ordinating vegetable growing in school gardens in Swaziland. Thought I knew everything but actually nothing about African conditions and then came back and started working with my father in the mid-70s. We selected better strains of Cox with better colour and planted them on MM106 very close together at 15ft x 10ft. We knew we had to replace the older orchards, so the first plantings were grubbed out in the early 1980s.

You are growing what the market wants?

Yes, I grow what the supermarkets want. Nowadays I am also planting for the retail trade and farmers' markets, but you only need a small acreage to supply the farmers markets.

When did you introduce cherries?

My father wanted to introduce cherries just before he retired. Then they had a rootstock called Colt, and prior to that they had F12/1 which were very vigorous. It took 7-8 years for the tree to crop because they were so vigorous. The trees were so huge that you needed to employ somebody with a gun to keep the starlings off. Cherry growing became uneconomic. Health and Safety did not like us going up these huge ladders to pick the cherries. It is only in the last 10-15 years we have these new rootstocks from America called Gisela and now we can grow small trees and put a net over the orchard and they are very prolific cropping. If necessary we can put a lattice rain cover over the trees to prevent the cherries from splitting. Cherries are coming back into fashion and they are very easy to sell.

Farming has been incredibly successful feeding the world population even though it is increasing by a billion every few years. Farming has kept up with it in terms of research and technology. If food prices get cheaper and cheaper, farmers will need to find alternate ways of producing it but less people will go hungry. The issue is if we can reduce our dependence on fossil fuels. Organic is one way of doing it. I think genetically modified food, if we can overcome our resistance to it, is the future. If we can have nitrogen-fixing bacteria on our wheat crops that would be huge breakthrough because a lot of fertilizers depend on oil for their production. If we could breed disease-resistant crops then we could all go organic and use less oil-based plant protection products.

Tom Maynard at Maynards Fruit Farm, Ticehurst

Maynards Fruit farm just outside Ticehurst was planted in the 1950s and was one of the first Pick-Your-Own orchards in Sussex. Today there are apples, pears, plums, cherries, cobnuts and soft fruit grown for the local market rather than for the supermarkets. Seventy acres of blackcurrants are also grown under contract for Ribena.

We are at Ticehurst here and at the top of the farm it is about 400 hundred feet above sea level. My father was brought up on a farm down in Stonegate which was in a valley and in a frost pocket and tended to get frosted quite frequently. His father was growing apples down there since 1930. My father's experience of getting frosted led him not to take over his father's farm but to get a site where his main criterion was to be frost-free and so that is why we ended up on a hill.

What has made the farm successful is the geography, the altitude combined with a reasonable enough soil. Other farms had allowed people to come in and glean what they could, but my father wanted to do it the other way around. We will plant for the public to come and pick your own and we will advertise for this, which was a new kind of thing. In the early 60s it was the time when people were getting their first car and deep freezers. People would pick for economy. That really took off and was highly profitable and we had cars parked right down to the village. That was on soft fruit and plums. For the first few years of the apple orchard, the entire crop was sold as Pick Your Own. In those days people would buy 60 to 70lbs of apples to store under their beds or in their garages or whatever, so they planted Bramleys to suit that market.

But you could get a good price?

We could get a good price... The point about cheaper fruit was that early in the days of PYO price was a major reason for people to get their fruit from us. It was not cheaper fruit in the supermarkets that caused a decline in our sales, but the fact that with increasing affluence people were not motivated to pick fruit in order to save money. We were probably slow to see that the main reason for people coming out was for a leisure activity, and that has to be marketed differently.

What types of soft fruit were there?

Well, there were strawberries, raspberries, gooseberries a bit of rhubarb and currants. The blackcurrants were another bit of the enterprise because from the beginning they were growing blackcurrants for Ribena under contract.

Was this at a separate site?

Yes it was a separate site and it was not open for the public. For a couple of weeks a year it needed to employ 20-30 casual pickers, then in the mid- to late-70s machine picking came in. Three growers combined together to buy a picking machine and

Tom Maynard and his father David grow cobnuts and a whole range of other fruits

subsequently that has become the biggest part of the business. Now the blackcurrants are not grown on this site and I rent about 70 acres of land off the site.

Returning to the apples...what varieties were planted?
There was just Cox, Golden Delicious and Bramley all on M9. Then in 1980, my father planted five acres or so of Katy apples. At the time it looked a winner in terms of its cropping and its qualities and it has a very nice flavour and crispness. A lot of good qualities. It had a particular market niche.

What others crops do you grow in the orchard?
Cherries have always been very popular with the consumer but production problems included frost, rain causing cracking (of the cherries), birds and more. The arrival of Colt rootstock in the 70s and now the Gisela rootstock allows smaller trees which can be picked from the ground and netted, and produce quality fruit from soils such as ours.

Cobnuts are grown on the poorest soils on the farm, and most of the pruning is with a hedgecutter. They are for harvesting and selling fresh in August and September and are extremely popular with a tiny minority of the public who will pay good prices and buy big quantities. However, picking costs are high and when the niche market is satisfied it is hard to sell any more, and we are sometimes left with substantial unsold quantities. Squirrels are a major problem, sometimes taking half of the crop.

Are you growing any new varieties of apple with disease resistance characteristics?
There are many new apples coming onto the world market with the ability to store for 10 months, and potentially good flavour and texture. Disease resistance, particularly to scab, is starting to get better. This will squeeze production on our scale even more, and the niche variety I am trying on a very small scale is Pinova which has Cox as a grandparent, has very good flavour from December, and is reasonably resistant to scab and mildew.

And in the future?
The apricots we have planted have proven the potential market for farm shop and PYO sales, and I think we have the varieties to suit the climate but it is early days. So selling local where we can and growing on a scale to meet that market remains our general strategy.

Matthew Wilson, EH Wilson and Sons at Oakwood Farm, Robertsbridge.

Oakwood Farm is a successful organic orchard that has been cultivated since 1951 by the Wilson family. It lies in an undulating wooded landscape near Robertsbridge. The orchard rows are planted with a wide range of apple pear and plum varieties many of which show resistance to the diseases which fruit are prone to such as scab and canker. This characteristic is essential in the organic production of top fruit. Sales are predominately direct to wholesalers followed by box schemes, retailers, farm shops and farmers markets. Oakwood Farm apple and pear juice is also sold across Sussex.

Oakwood Farm is an 80-acre organic orchard that has been cultivated since 1951 by the Wilson family. It lies in an undulating wooded landscape six miles from Robertsbridge. The orchard rows are planted with a wide range of apple pear and plum varieties many of which show resistance to the diseases which fruit are prone to such as scab and canker. There has been gradual experimentation with varieties which were said to be disease resistant such as Adams Pearmain and Court Pendu Plat before bulk planting across the orchard. There are also modern varieties such as Rubin which have been bred for their disease resistance. This characteristic is essential in organic production since there are only a limited range of materials which can be used in the control of disease. There is also an emphasis on maintaining a natural environment of flowering plants within the orchard to attract predators of the different types of aphid and the use of seaweed and iron as a foliar feed.

The Oakwood Farm site had originally been pasture, with 6 acres of hops, and then in 1946 it had been bought by Raymond Wickham, a Kent fruit grower. The orchard was first planted with apples, pears and plums grafted on vigorous rootstocks, though because there was insufficient protection against rabbits the trees were badly damaged in the first winter.

In 1951 the orchard with its damaged trees was bought by E.W. Wilson, the grandfather of Matthew. He had cultivated 30 acres of apples at Yalding and glasshouse crops at Billingshurst but by profession he was a banana merchant. The early plantings were of Cox's Orange Pippin together with Laxton's Fortune, Worcester Pearmain, Tydemans Early and Lord Lambourne. There were also six acres of cooking apples as well as plums, pears and 12 acres of cherries. Some of the orchards were also planted with loganberries or blackberries but this practice was soon abandoned because it caused access problems for tractors and the plants got entangled with the rows of apples.

There were five permanent staff as well as part-timers and seasonal packhouse staff. During the season 1000 boxes a day of Worcester Pearmains were transported by Turks Transport of Benenden to the Liverpool market wholesaler Mark Revil.

During the 1970s the original orchard trees were gradually grubbed out as there were too many early varieties which were not selling as European imports undermined the English market. The trees were also being replaced because they had come to the end of their productive life. There were new plantings of Cox, Egremont Russet, Spartan and Fiesta on semi-dwarfing rootstocks and in the 1990s Jonagold, Gala and Braeburn were introduced.

Matthew Wilson is a leading Sussex organic grower

During the 1950s and 1960s there was an intensive use of herbicides, fungicides and pesticides, though lead arsenate and mercury were soon replaced by more modern chemicals such as Captan, Triazole fungicides and organophosphorous insecticides. The overuse of these chemicals led to an increase in resistance by insect populations and certain growth disorders.

The high cost of chemicals was one of the motivations behind the movement towards organic production together with the understanding that their use to get complete control of insect populations destroyed both the pest population and its predator. The change to organic production was also influenced by the fact that organic crops, while difficult to grow, are easier to market and sell, especially today with the demand for sustainable food.

Sales are predominately direct to wholesalers, then, in terms of volume,s to box schemes, retailers, farm shops and farmers markets. Supermarkets are thought to be too selective of variety, size and colour and are not used. Smaller grade apple and pears are made into juice, bottled and distributed throughout Sussex under the Oakwood Farm label. In the last few years there has been an increase in the number of different crops being grown at Oakwood Farm, with the cultivation of walnuts, late-flowering varieties of apricot, almonds, quinces, cobnuts and medlars.

Tony Eales at Tendrings Fruit Farm, Magham Down

Tony Eales had worked in dairy farming ever since he came out of the army. He bought the orchards at Magham Down in 1975 when there were 25 different varieties of apples, with pears, plums and greengages. Most of the fruit was sent to Hellingly station for the Covent Garden market or sold at a shop in Hailsham. Then as local labour supplies dried up and the marketing demands of the supermarkets increased, Tony was convinced that at the age of 60 it was not worth investing further in the orchards. Today Tendrings Fruit Farm is run on a smaller scale with a farm shop and orchards alive with birdlife and with grazing sheep under the trees.

Tony Eales sells a wide range of apples from his own farm shop in his orchard

I had been in dairy farming ever since I came out of the army in 1947 and did not come into orchard growing until 1975, so my knowledge of apple growing was limited. I was working in the marketing end and left the practical work to the original permanent staff (at Magham Down Farm) who had spent their entire working lives here.
Before the war and afterwards apple trees were planted for a continuous supply of

different varieties of apples and plums through the season from July to November. Half the area of the orchard would have been picked by mid-August and dispatched by Hellingly station for Covent Garden market. There was of course no cold storage in those days. There were 25 different varieties of apples, 4 pears, 8 plums and greengages. When I took over, there was an existing farm shop. I then established a shop in Hailsham high street, but eventually the shop rents and rates became too high, whilst the fruit prices stayed the same.

With my advancing years and my two sons having other careers, the future of the orchards seemed to be at risk. To run this sort of business you have to supply the supermarkets, or you are into supplying outlets such as farmers' markets and local shops which cannot absorb huge quantities of fruit. However, supermarkets now insist that their fruit is graded and packed in a central pack house, which is forced to dispose of less than perfect fruit at prices which are not profitable for the grower. I used to supply an agent who acted for supermarkets and we graded and packed here at the farm with 20-30 pickers and eight staff in the pack house. We sent away sometimes three tonnes of apples in a week to Sainsbury's and Somerfield's. When they decided they would only accept suppliers from designated packing stations, we could not continue. At about that time the local labour supplies dried up, and whereas in 1975 there were plenty of ladies who were prepared to come and pick fruit for 6 weeks, they slowly disappeared. They preferred work that lasted throughout the year, they're just not here anymore.

In 1987 we suffered a disaster from the hurricane which cost me tens of thousands of pounds; there were still apples on the trees, field after field just blew over, which then took me 10 years to clear and refill. We had some very tall wind breaks on the farm which also blew over and fell on everything else. Financially it was a disaster, which at 60 made me question whether I should replant or not.

I reached the stage when I was 63 (83 now) when I realized that if I planted trees they would cost around £10,000 an acre and in very few years I would be drawing a pension. I wondered; do I really need to do this? Costs had gone up whilst returns were static. I invested in a coldstore for the fruit which enabled me to keep apples through to the spring. Lower grade fruit was sold for juicing.

*The orchards are a bit small now?**

Yes, but by running a farm shop and selling as much as we could direct to the public, it became more profitable. There has been a problem and I could not see a progression, the trade through the farm shop gradually reduced as people went to the supermarkets. However, despite the forgoing we like living here and we therefore carry on, and maybe one day young trees in a small intensive orchard will be established.

Eric and Janet Rowlands (with grandchild Sam) grow organic apples in part to safeguard the purity of the water reservoir at the bottom of their valley

My overriding aim is to provide for the future a wildlife friendly park-like habitat. We already have very prolific bird life on the farm. (See chapter 11).

NB: Much of Tony Eale's 50 acres of orchard are used as grazing for sheep, and provide a rich wildlife habitat; sometimes the apples are picked, but only 4 acres are managed.

Eric and Janet Rowlands at Great Sanders Fruit Farm, Cripps Corner

Great Sanders Fruit Farm at Cripps Corner has been planted with fruit trees since the 1920s. The original trees were grown on vigorous M2 rootstocks which grew into large trees. Eric Rowlands has worked at the farm since 1973 and became the owner of the 25 hectare-orchard five years ago.

Originally the orchard formed part of the 566 hectare estate owned by Southern Water which also included Brede Woods, which has recently been sold to the Woodland Trust. The orchard lies on silty-clay soils of the Tunbridge Wells beds which Eric Rowlands does not consider very good quality. Great Sanders is an organic orchard and part of the decision to cultivate the orchard organically is to prevent pollutants running into Powdermill Reservoir at the bottom of the valley.

Today the orchard is planted with Cox, Bramley, Spartan, Chivers Delight, Egremont Russet, Tydemans Early, Elstar and Discovery on dwarfing rootstocks such as M26, MM106 and M9. Eric has been experimenting with a new Italian variety Modi, a crisp red apple, which was said to be scab and mildew resistant. However under Sussex conditions Modi has proved to be susceptible to both. In the future Eric will be planting more early varieties of apple like Discovery, of which there is a shortage in the orchard, and planting Crimson Crisp a recent introduction from America with resistance to the fungal disease scab.

The orchards are sprayed with sulphur during the early summer months to prevent scab, and the biological control *Baccillus thuringiensis* for caterpillar pests. Pyrethrum is used for Blossom Weevil. In past years, Tesco was one of the main markets for the apples grown at Great Sanders Fruit Farm, but increasing quality demands and reduced prices has ended that contract. Eric now sells to companies specialising in organic fruit or apple processing companies. Riverford Organics and Abel & Cole buy much of his fruit and Fourayes of Sittingbourne process the apples into pies or organic babyfood. There is also a market for the fruit with Roy Cook for juicing and Aspall cider.

Jim Buckland at West Dean Gardens

When Jim Buckland and his wife Sarah came to West Dean in 1991 the gardens were run down and partly derelict. He proposed that the walled garden should be redeveloped in the spirit of how it would have looked in the early twentieth century. Today with the support of the Edward James Foundation, West Dean Gardens contains an immaculate kitchen garden and an orchard of a hundred fruit trees with many from Sussex. The historically important glasshouses have been renovated and contain vineries and trained peaches. West Dean is well known for its exuberant Chilli festival, Tomato Day and Apple Affair when visitors walk through the gardens set in a landscape of parkland and farmland.

When we came here the gardens were generally run down. The walled garden itself was derelict, especially the glasshouses with the exception of the fruit garden which was a garden centre with a modern Venlo glasshouse. The Trustees decided to shut the garden centre prior to my taking the job as garden manager.

We proposed that we redevelop the walled garden on the lines of how it would have looked in the early twentieth century, that being the period when it would have been at its peak. Edward James's father, William bought the house and estate in 1890 and being a very wealthy man poured money into it. They were great socialites and Edward VII was a frequent visitor here.

We never set out to do a facsimile edition. The aim was to capture the spirit of the era and recreate the gardens in a manner that would have been recognisable to our Victorian and Edwardian forebears but with a large number of caveats.

There was an existing kitchen garden?
Yes, we imposed a four-quarter design which was not historically accurate but was more attractive. In the original there would have been large beds with paths across. The glasshouses had not been touched for 35 years so were in a parlous state. But once the vegetation had been cleared away we could see that most did not have panes of glass but were fundamentally sound... we replaced about 10 per cent of the timber... We have three vineries and a peach and nectarine house. The glasshouse range is the best example of a homogenous glasshouse range as they are all built by Foster and Pearson between 1890 and 1900. They are kept in good condition and painted every four years. Maintaining the glasshouse is the second most expensive part of our budget. The heating system was shot to pieces so we did replace that.

Jim Buckland at West Dean masterminds the estate's popular Apple Day and tends a beautiful orchard where there are lots of Sussex apples

The trained fruit, were they chosen to be part of the Victorian gardens?
The collection is based on having a good comprehensive collection that does have a historic basis. We do have some archival material from the turn of the century, and if

a plant is mentioned, available and worth growing we would grow it... but we would not grow it just because it was mentioned... it might be a poor variety.

The intention has been to grow as many Sussex varieties (of apple) as we could source at the time...which meant going to [the National Fruit Collection] Brogdale, and then asked Frank P. Matthews to bud and grow them on for us. They produced four of each variety and then they kept two.

The trained fruit – does it reflect what was grown before?
It does in terms of the walls. Espaliers and cordons you would have seen, but not the intricate cordons or bowl-shaped pears which derive from France and the Lorette system of pruning. Growing the half standards (apple trees) in the orchard is more attractive and feels more like a traditional orchard. Historically you would have seen fans, espaliers, cordons but probably not goblets and pyramids.... They have a nice quality when they are bare, especially when it snows.

What about Apple Day?
The Apple Day [called the Apple Affair at West Dean] is partly to attract visitors and partly to educate people about growing. We are quite commercial about it. It is as much about generating publicity as pulling loads of people. Would we do it if we did not make any money... possibly not. Apple Day is a national thing promoted by Common Ground and we latched onto it.

It has grown over the years but has not changed very much, but people seem to like it and out of our smaller events it is consistently successful. We have got the local schools involved for the last ten years which has been a great success...both in terms of community involvement and it gives it a different feel because we are not a hugely kiddy place a lot of the time. The Apple Affair is always my favourite show because of the feel of it and people sense that the nights are drawing in and it is the last time they will get out.

9. Growers

9

Professional fruit growers in Sussex

High-tech orcharding: today's fruit growers can sign up to receive weather alerts via their mobile phones

In Britain as a whole, and in Sussex too, commercial apple-growing is on the increase, according to Adrian Barlow, head of the industry marketing organisation English Apples and Pears. The modest upsurge in the past few years reflects "public interest in local supply and concern over climate change," he says. After decades of decline, the new millennium seems to have marked the beginning of resurgence in apple cultivation for the marketplace.

Although about 70 per cent of the UK's eating apples are imported, most supermarkets now stock British apples much of the year, and some supermarkets, such as Waitrose and the Co-op, grow apples on their own farms.

An apple is an apple is an apple? What best to grow?

But growing apples commercially for Britain's highly competitive, globalised market remains a challenge, and chief among growers' concerns is deciding which apple varieties to grow. In the post-war years, commercial apple producers were advised to concentrate on growing Cox, alongside Bramley's Seedling to satisfy demand for cooking apples. By the 1980s, Discovery, a twentieth-century variety bred in Essex, was also a commercial favourite as an early variety, but its short shelf life has fuelled its sharp decline. Two older English varieties, Worcester Pearmain and Egremont Russet, with Sussex roots, still sometimes appear on supermarket shelves, but increasingly compete with new, heavily branded varieties.

Since the early 1990s, Braeburn and Gala, two varieties with New Zealand origins, have proved especially popular with British consumers, and are now outselling Cox by a substantial margin. Year-round marketing has contributed to the newcomers' popularity, as has their attractive colour and long shelf life.

To compete in the beauty contest, English growers responded by adopting newer Cox varieties that arguably don't taste as good as the old variety but are redder, bigger and less prone to cracking, yet the sales of Cox are still falling year on year. Market analysts say that consumers now want a sweeter apple, rather than the mix of sweetness and tartness that typifies the traditional English apple, and that the market for Cox is likely to continue to decline.

So British growers have begun growing Gala and Braeburn themselves, now that warmer autumns make growing these varieties feasible, especially in southern counties such as Sussex – although both varieties are very prone to scab in our climate. Still, with fungicidal sprays to hand, British growers produced a record crop of Gala in the 2011/12 season, with sales up some 14 per cent from the previous year. "British-grown Gala is taking an ever greater share of the Gala market," says Barlow.

Pink Lady is one of the new "Club" variety on the market, with a "trademark brand" image. Growers have to apply to grow the variety, meet high specifications, and pay a levy to the Club variety licence holder. Pink Lady, originally bred in Australia, cannot be grown in chilly Britain, but several other Club varieties can be, including Jazz, Kanzi, Cameo and Rubens. Jazz from New Zealand is a cross between Gala and Braeburn, as is Kanzi, also from New Zealand. Cameo is from Washington state in the US, while Rubens is a cross between Gala and Elstar, bred in Italy.

Bramley remains Britain's premier cooking apple, but only because growing sharper-tasting apples for pies, cakes and purees is a British speciality. Sales of Bramleys are falling, year on year, as consumers reject cooking apples as being too time-consuming, although food processors now often use Bramleys in ready-made desserts.

In Sussex, C&M Ramus & Sons at Iden is the largest commercial producer of top fruit. Targeting the wholesale market, they grow chiefly Cox and Russet as dessert apples, and Bramley – the "apples which people look for as being English," says Charles Ramus (see interviews in chapter 8). To grow the fruit economically, he and his father Nicholas have cut labour costs by mechanically pruning the Cox's, using a machine originally devised for French vineyards, backed up by periodic hand pruning. But it's a constant challenge to get enough light into the trees at just the right time of year to maximise the colour and size of each fruit, and people have to work in the orchards all summer, thinning and removing slightly damaged fruit to ensure perfection for the remainder. "Cox is the most pernickety, horrible tree to grow," says Charles.

When their Cox trees on a particular dwarfing rootstock reach what's deemed the advanced age of 30, they're replaced by newer Cox clones – for instance, the Ramuses have tried a new variant, Cox sport[1] Lavera, which has better colour and bigger size "but the taste is not so good," says Nicholas. Still, they reason that if most supermarket suppliers are switching to Gala, they're better off sticking to Cox, with its recognised brand name, while adding a small number of Red Windsor trees to the orchard to produce an early crop. Judging the market is a constant gamble. "We have a trial crop of 500 Braeburn," explains Nicholas, "to see how they grow compared to Cox."

The timing of harvests also influences the choice of varieties. As the early Cox come out of store to go to market, the Braeburn, picked in October, can take their place in the storehouse. The Ramuses have also planted Estival, which is ready for harvest much earlier in the season at the end of August, "because labour is going to become a really massive issue". Nicholas explains: "With the trouble our little country has got into it is not going to be seen as this pot of gold to come and work in, and universities are becoming more and more fussy about when their students can work, so we don't want too much after the first week of October."

"So that is why Red Windsor would fit in as we reduce the amount of Cox, and grow a small amount of Braeburn, because I do not want to have to get pickers late in the season." It's a question of getting the mix and balance right, Nicholas emphasises. "There are always fad apples like Rubens, Kanzi, Cameo, Jazz. One will be really successful, but which one?"

The Bramleys, formerly on big, standard trees but these days mostly on much smaller bush-trees, are now harvested in August instead of the end of September. "Normally we have finished the harvest of the Bramleys before we have started the Cox," Charles explains. The Bramleys on the dwarfing rootstock are ripening a month earlier than before, growing bigger, and must then be picked quickly before they ripen further and lose their trademark greenness – "people do not like any red on the Bramley, they like them green, because people shop with their eyes," he adds.

Graham Love at Greenways Fruit Farm in Herstmonceux grows Gala, Braeburn, Elstar and Jonagold for the supermarkets, as well as a wider range of newer varieties for the retail greengrocers and farmers' markets. Some years ago, he and his father planted modern strains of Cox with its better colour, but explains that Cox needs a lot of costly management, so "people are not planting it at the moment. Everybody is going to Gala and Braeburn and various new crosses of these varieties."

Unfortunately, he concedes, "Gala is very prone to canker and scab and Braeburn get scab badly. But on the other hand they have four times the cropping of Cox...so even if you get a lower return on Braeburn and Gala there is a better return per acre." Another problem with Gala is that "it is a small apple which goes a bit soft and is easy to bruise. Braeburn is a hard apple which is late to mature...and we'll still be picking it in November." Graham explains that there are lots of crosses between Gala and Braeburn, and new exotic varieties such as Jazz, Fuji, Pinova, Cameo and Evalina. "Every packhouse is patenting these varieties which controls who can plant them. So you have to sign a contract that prevents you selling it outside the group." Graham also grows two English varieties, Discovery, which he says is "quite disease-resistant, and Charles Ross, as well as the Canadian Spartan, which gets a lot of canker. We do grow Elstar, a popular North European apple, and the American variety Jonagold. Some of these modern varieties are fairly disease-resistant though Braeburn does get mildew."

At Ticehurst, Tom Maynard of Maynards Fruit Farm grows Cox, Golden Delicious, Katy and Bramley, and is trialing a new variety, Pinova, with some resistance to scab and canker, as are several other Sussex growers.

In the 1960s, his father planted Cox, Golden Delicious and Bramley, and in the 1980s added five acres of Katy apples – a Swedish cross between James Grieve and Worcester Pearmain, introduced to England by a Kent fruit grower in 1966. "At the time it looked a winner in terms of its cropping and its qualities, and it has a very nice flavour and cripsness. A lot of good qualities. It had a particular market niche." But sending them off to supermarkets proved problematic, as packing specificiations became ever more stringent and the fruit bruised easily. "What we wanted to do was grow to a market niche, growing blackcurrant for Ribena and apples for the local population with our pick-your-own scheme. On our soils and situation, we knew we could not compete with more efficient growers on a large scale." Tom adds: "We planted Estival apples because we can market them from the tree, and they keep their superb flavour and texture for several weeks, unlike most other August/September apples. Yield is good but bruising is a major problem. They are also very susceptible to scab."

The future remains uncertain. "There are many new apples coming onto the world market with the ability to store for ten months, and potentially good flavour and

texture," Tom explains. "Disease resistance, particularly to scab, is starting to get better. This will squeeze production on our scale even more, and the niche variety I am trying on a very small scale is Pinova, which has Cox as a grandparent, has very good flavour from December, and is reasonably resistant to scab and mildew."

New varieties on trial: Sussex grower Tom Maynard is experimenting with a new German apple variety, Pinova, said to be less vulnerable to fungal diseases

Meanwhile, organic grower Matthew Wilson, of Oakwood Farm at Robertsbridge, grows Cox, Egremont Russet, Spartan, and Fiesta (or Red Pippin) as well as many of the older varieties of apple which are less prone to fungal diseases such as Adams Pearmain and Ribston Pippin. Some of the more scab-prone varieties in the orchard are gradually being replaced by varieties more suited to organic production. Matthew is also trying Rubin, from Czechoslovakia, and has planted this year three new scab-resistant varieties, Santana, Saturn and Rajka. He has also recently 'topworked' some of the orchard tree with Sussex apple varieties.

Matthew Wilson inspects the Sussex apple tree scions grafted onto an established tree in the practice known as top-working

Keeping pests and diseases at bay

Pests and diseases are a notorious challenge for commercial apple growers. Pressure is intensified by the premium placed on high yields, low costs and perfect fruits by today's supermarkets, where 80 per cent of all apples purchased in England are sold. "Fifty years ago, most fruit of reasonable condition would have a saleable value, but today only Class 1 quality standard has any chance of realising a value above the costs incurred from harvesting to point of sale," argues the author of "The English Apple Man" website. Heavily marketed varieties popular with consumers are often prone to diseases such as scab and mildew, and modern commercial orchards, with closely packed bush-trees in large monoculture plantations, creates the perfect conditions for outbreaks of pests and infections. Although the use of pesticides and chemicals has evolved considerably since, say the mid-twentieth century (see chapter 7 for a brief history), reliance on a battery of pesticide sprays remains standard practice in most professional enterprises.

Scabby apple: this fungal condition doesn't affect the fruit's flavour or nutritional qualities, but even a speck makes an apple virtually unsellable

The mainstream commercial fruit industry makes use of a host of newer pesticides, including organophosphates and the neonicotinoids, recently linked to the decline in bumblebees and honeybees,[2] under relentless pressure to produce a high volume of perfect-looking apples for supermarkets. Today, most commercial producers repeatedly spray their apples with an array of different pesticides over a seasonal cycle to reduce both insect and fungal pests, such as aphids, mildew and scab. Intensive pruning removes dead or decaying wood, and mowing and spraying with herbicides keeps the orchard floor clear of vegetation – all in all, making the modern orchard as a whole of little value for wildlife. In recent years most growers have adopted a reactive policy of spraying as part of an "integrated pest management" system. Instead of spraying routinely at particular times of year, growers attempt to monitor pest populations and time their applications of chemicals more precisely to maximise the impact on the pests, and to reduce their use of costly chemicals.

Graham Love at Greenways Fruit Farm, Herstmonceux, explains his techniques for dealing with one pest: "Codling moth is the main caterpillar that will eat into the fruit. We use pheromone traps to monitor the codling population. The female gives off a pheromone when she is ready to mate. We use an artificial pheromone to attract males and when they are found in large numbers you know that mating is occurring. You can then time the application of the pesticide to catch the caterpillars as they emerge from the egg when they are at their most vulnerable. You can then use the least amount of pesticide to control the pest."

When the apple buds burst in March/April, Graham and his team put "protective fungicides on the opening foliage." He explains: "if you can protect the foliage from scab in early spring you can save on sprays later as the scab has not got established. This means that less products are sprayed onto the actual fruit."

A few decades ago, Graham's father used tar oil, copper, mercury, sulphur, arsenic, and a host of persistant organochlorine pesticides including DDT, dieldrin and aldrin. "Now we are so environmentally conscious that we use ten insecticides to kill specific pests but not beneficials [insects and other natural predators]. Modern growers are concerned to maintain a balance within the orchard."

"Now all fruits for supermarkets are tested for residues and I am audited each year. If any chemicals are found which should not be there, they will not market the crop. Before we could apply chemicals before storage but after picking to prevent it rotting. Now we have to apply these materials while the fruit is on the tree, using up energy and leaving chemicals in the environment." He wonders what will happen if there's a wet September and damp fruit goes into the coldstore. And as summers grow warmer, as climate change kicks in, insect pests will be more of a problem, Graham predicts, and fungal diseases less so. "Recently, wasps have been an incredible problem. We spend all this money on insecticides then ten days before harvest wasps come in and do incredible damage – they go for certain varieties of apples, pears and cherries. We cannot use insecticides on the wasps so close to the harvest as it would leave residues on the fruit."

Herbicides play a crucial role in controlling the growth of weeds and kill grass growing under the dense plantations of dwarf bush-style apple trees. "Grass is a hungry feeder so it forces the tree roots to grow downwards below the best soil which is in the top two inches. Once herbicides came in there was a huge increase in cropping and fruit size," says Graham, "because the whole of the soil became available to the apple roots to absorb moisture and nutrients."

In Britain, few growers have opted for a fully organic system, and imports are still satisfying most of the demand. Organic apple production on a commercial scale is not easy. But Sussex is home to one of the leading organic top fruit growers in the country, Matthew Wilson at Oakwood Farm, Robertsbridge. His family-run business began organic production in 2000. The high cost of chemicals encouraged the move towards organic production, together with the gradual understanding that the increased use of chemicals in an attempt to completely wipe out a pest species destroyed beneficial insects too, including pollinating bumblebees as well as the pests' natural predators.

He and his wife Carol now have 40 acres given over to organic fruit, including 12,000 apple trees. Their over-riding philosophy is to "try to look after nature and it will look after you," by encouraging natural predators in the orchard that can tackle many of the pests. The Wilsons simply mow the "tumbledown" grasslands beneath the trees, enabling orchids, violets and wild strawberries to flourish, along with earthworms and other beneficial invertebrates. Refuges are put out round the orchard for earwigs and ladybirds which feed on aphid pests. To reduce the breeding of codling moths, the Wilsons use an ingenious system of pheromone traps, called Exosex. Dispensers placed throughout the orchard – stocked with powder scented with the pheromone of the female codling moth – attract males which then get covered in the electrostatically charged powder. Mating is disrupted because the male moths can no longer detect the female moths, and males covered in the smelly powder also lure away newly emerged male moths.

To help to reduce caterpillar damage, Matthew also uses other "biological control" techniques, such as applications of a specially formulated solution of a bacterium *Bacillus thuringiensis*, and codling granulosus virus – demanding techniques that need to be precisely timed to coincide with stages in the lifecycle of the caterpillars.

Fungicidal sprays for scab are largely confined to sulphur mixed with a little copper at a rate approved by the Soil Association. The fruit trees are sprayed with sulphur early in the spring when diseases such as scab are most likely to spread. "We find that, as far as scab is concerned, it's important to mix your varieties up, breaking up large areas of monoculture." The emphasis is on growing the fruit trees in an ecologically diverse environment with a healthy soil and beneficial wildlife. When the vegetation beneath the trees is mown, cuttings are left to rot and return nutrients to the soil. The Wilsons manure the orchard soil and spray the trees every two weeks with a liquid feed derived from seaweed.

Tony Eales at Tendrings Fruit Farm at Magham Down near Herstmonceux more or less stopped spraying his apples some years ago, when he hit retirement age and decided to run the orchard less intensively. Although he hasn't planted any trees for about 40 years, many of the trees, on semi-dwarfing rootstock, are still productive, and he has some 20 varieties of apples to sell at his farm shop, and for juicing. His orchard doesn't qualify for organic status, as he uses copper sulphate and a fungicide once in early spring, as well as herbicides to control the vegetation under the trees, but he can sell his apples as "insecticide-free".

"To grow apples for a supermarket you have to keep them blemish-free, so one had to spray against insect pests, fungi and mildew. I gave up this intensive use as soon as I could and went down the beneficial insect road, realising that it was ridiculous that we were killing off the predators." Tony reckons that because he uses few chemicals his orchard benefits from an "overwhelming number" of beneficial predatory insects that keep the pests down as a result. Some of Tony's varieties are vulnerable to scab, but he notes that others, such as the early variety Beauty of Bath, never get it.

His apples aren't perfect by supermarket standards, but nearby Middle Farm takes a weekly supply (from Tony's coldstore) to make fresh juice to sell. He also sells his apples – including Bramley's, Bleinham Orange, Cox's Orange Pippin and Russet – at his own farm shop, as well as honey made from hives in his orchard. He says his Cox's are the old traditional sort: "supermarket producers use other modern clones that produce apples that look prettier and grow bigger but they don't taste the same."

Shrinking orchards, dwarfing rootstocks

Commercial orchards today are not usually stocked with full-sized apple trees, and growers have to decide how they want to grow their apples. Most are planted with apple varieties grafted onto dwarfing rootstocks that produce very small bush-like trees placed very close together – at planting densities of up to 3,000 trees per hectare. Before this "modernisation", orchards typically consisted of widely planted large so-called "standard" trees, planted at a density of no more than 100 trees per hectare. The switch to dwarfing rootstocks began in the early part of the twentieth century, but became widespread from the 1950s onwards. Trees on dwarfing rootstocks produce more fruit more quickly than standard trees, and are easier to spray with chemicals and to harvest with no needs for ladders.

Most of the important earlier rootstocks came from the East Malling Research Station in Kent. Each stock was given a number preceded by an M for East Malling. The dwarfing rootstock M9 is grafted with an apple variety that grows to a height of no more than six or eight feet, and is planted extensively in commercial orchards. It produces an early fruit-bearing tree within five years, but it has a weak root system

and needs permanent support. Other rootstocks used by commercial growers include M26 which produces somewhat larger, more robust trees growing eight to ten feet in height. Dwarfing trees do not have the same vigour as their larger counterparts, but may crop more reliably. They are pruned each year to maintain the desired shape and to allow sunlight to reach all the fruit to ensure a good colour.

Commercial orchards today: serried ranks of apple trees on dwarfing rootstock in C&M Ramus's intensive orchard system

In 1917 the East Malling Research Station joined forces with the John Innes Institute at Merton to breed some aphid-resistant rootstocks. One of the outcomes of their work is the present day MM106, a semi-dwarfing rootstock, producing a tree of between 9 and 11 feet tall, that can grow in a wide variety of soil conditions. The orchards of C&M Ramus and Sons in Sussex use a combination of rootstocks: an M9 interstock grafted onto an MM106 root system. M9 provides the necessary dwarfing characteristics for an intensively managed orchard while MM106 provides a stronger rootsystem. The apple variety is then grafted on the M9 interstock. (For an image showing how rootstocks determine the eventual size of a tree see chapter 12.)

And then there's the weather…

As well as changing fashions in apple varieties, apple growers have to contend with the vagaries of wind and weather, and a relentless drive to increase productivity in the face of growing costs. The hurricane of 1987 blew over hundreds of Sussex apple trees still laden with fruit, spelling the end for at least one commercial orchard, the Ashdown Land Company near Horsted Keynes, and triggering major change at Ramus & Sons, as big old trees were destroyed. "Three-quarters of the young standard orchards were gone overnight by the storm, and half of the established ones," says Charles Ramus. "After the hurricane, young bush-type trees were put in to get production back quickly." At Magham Down, Tony Eales lost "tens of thousands of pounds" and took ten years to clear and replace all the trees that had been felled by the wind. The ill winds of the economy also continue to intensify. Labour costs have risen dramatically, and availability of seasonal labour has declined, while the returns from sales remain static; meanwhile, the demand for uniform and unblemished fruit continues to intensify.

Finding markets, selling locally

The bulk of Sussex commercial orchard production is sold via wholesalers that supply supermarkets. Yet it can be very difficult for small-scale growers to meet the supermarkets' stringent requirements, and many apples get discarded in the process – rejected for being too small, too big, too russeted or too unevenly coloured. As a result, many of today's growers including organic grower Matthew Wilson, have abandoned supermarkets. He sends the bulk of his apples to wholesalers who supply independent shops and box schemes, and also makes various blends of apple juice with the visually less-perfect apples, which are widely sold across the southeast, including National Trust properties. Organic grower Eric Rowlands at Great Sanders Fruit Farm once sold his apples to Tesco, but now supplies companies specialising in organic produce such as Riverford Organics and Abel & Cole.

Graham Love in Herstmonceux grows a range of apples for supermarkets, as well as the retail trade and farmers' markets. "I take ripe Cox to the farmers' markets which have flavour, but people expect something which is crisp and hard and so they have to be re-educated. I have to explain that that the apples I supply to Tesco are picked two weeks earlier before the flavour develops in order to get shelf life and storability, which also suits the grower because they store well and bruise less."

Many fruit growers including those at Perryhill Orchard, Hartfield produce their own apple juice (an excellent way of making use of those less-than-perfect apples)

Both sets of apples may be from the same trees. One group – the right size and colour and blemish-free – go to the supermarket, but those going to farmers' markets around Sussex, while not technically so perfect, stay on the tree for two weeks longer and are picked on the day before the market is held. This extra maturity makes a key difference to flavour. Graham Love sees his fruit business moving away from its present 80 per cent reliance on supermarket supply to a more diverse range supplying local areas in Sussex. At Bill's Produce Store, the thriving deli-cum-restaurant in Lewes, it is possible to see baskets proudly marked "Graham Love's Discovery apples, grown in Herstmonceux".

Some fruit growers have their own farm shops where they can sell their apples, and usually their own juice as well, direct to customers throughout the apple season. From their farm shop, the Dench family at Ringden Farm, Flimwell, sell apple juice, which they began making in 1990, after a violent hailstorm marked the apples. Now, after harvest, their best apples go to farmers' markets and local shops, and the rest go for juice. The pomace, or dry pulp by-product, goes to pig food or to a worm farm in

No middleman: Grange Farm sells its produce from its own well-stocked farm shop at Funtington

Biddenden, where the worms turn it into compost. Tony Eales of Tendrings Fruit Farm once sold his apples at a greengrocer's shop he ran in Hailsham, but now sells to customers direct from an old wooden packing shed next to his house. He feeds his rotten apples and windfalls to his small flock of sheep which graze under the trees and produce up to 80 fat lambs through the year. Ivor Kiverstein took over Tullens Fruit Farm near Pulborough and sells much of his produce via farmers' markets and the farm shop. The farm now grows Cox, Worcester Pearmain, Bramley, Egremont Russet, Red Pippen, Lord Lambourne, James Grieve and Discovery as well as Blenheim Orange and Laxtons Superb. Incidentally this is the farm that was previously run by David Atkins, author of *The Cuckoo in June* (see chapter 7). At Perry Hill Farm, Hartfield, there is now a large selection of single variety apple juices, ciders and perry made from the farm's own orchards.

Some orchards allow visitors to pick their own fruit. One example is Redcoat Orchard at Catsfield, near Battle, an old orchard with beautiful views, and another is Grange Fruit Farm at Funtington, near Chichester, which has a range of top fruit and other produce including vegetables. At Stonehill Farm, Horam, you can pick your own plums as you can too at Durleighmarsh Fruit Farm, Rogate and at Maynards, Ticehurst, who also have "platts" (orchards) of cobnuts, as well as a range of soft fruits. Pick Your Own fruit farms emerged in the 1970s and fell out of favour in the 1990s, but are now experiencing a return to popularity as people come to appreciate seasonality, ripeness, flavour, freshness, and the appeal of the locally grown.

Pick your own: children from Ninfield Primary School visit Redcoat Orchard, where they were invited to taste the different apples and pick a bagful to take home

At Laines Organic Farm, Toos Jeuken, originally from the Netherlands, runs a thriving organic farm near Cuckfield in West Sussex, selling a wide range of vegetables for direct sale to the public. About twenty years ago she planted apple trees including Newton Wonder, Bramley's Seedling, Sunset, Golden Delicious and a russet. She sells her produce at local markets including the Lewes Friday market. Another Sussex enterprise markets both apples and eggs. Brambletye Fruit Farm is a 60-acre biodynamic orchard, Demeter certified, at the edge of the Ashdown forest at Forest Row in East Sussex with apples and soft fruit. A flock of 1600 free-range hens graze under the trees.

Sussex apple juice and cider revival

Many of the professional apple growers in Sussex are now producing their own apple juice, making use of the fruit which will not make the grade commercially as fresh produce and turning it into a delicious drink. Juices may be a single variety or a blend – so you can choose whether to go for a sweeter or a drier taste, for instance. Across the county in pubs, cafés, and local shops, it is increasingly possible to find bottles of apple juice made in Sussex where just a few years ago only the highly processed apple juice made from concentrate was available. Around the county, cider is also being produced, with locally grown apples.

John Batchelor makes traditional cider from cider apples grown in his own orchard at Maplehurst, and sold in the White Horse pub in the village. Richard Jupp collects feral apples from the downs and makes a distinctive cider aptly named Wildthing. At Pump Bottom Farm, Chichester, Julian Moores makes Appledram ciders, perry and apple juice and has a licenced cider house and restaurant on the farm. In Hastings,

On tap: John Batchelor sells his cider through his local pub

Andrew Etherton makes Battle cider from organic apples, while at Pennybridge Farm at Mayfield, Martin Clarke brews organic cider from apples crushed by a stone mill. At Gospel Green, a champagne-style fizzy cider is fermented to reach a heady 8.5% alcohol and the Fishlocks at Black Pig Orchard, Danehill make still and sparkling cider. In Bolney, Tom Stephens at Old Mill Fruit Farm is making apple juice and cider, with the brandname Wobblegate. At Middle Farm, Firle, visitors can taste an impressive variety of ciders from Sussex including Uncle Stoat's and Middle Farm's own brand Pookhill. Chris Hunt at Sedlescombe Vineyard near Robertsbridge makes organic apple wine, cider and and juices which are sold from its farm shop. Matthew Wilson in Robertsbridge makes organic apple juice and now cider too. This list is far from comprehensive, but gives a flavour of the many apple enterprises now thriving in Sussex.

Endnotes for Chapter 9

1 The sport of an apple is a variety or strain arising from a mutation on a previously named variety.

2 "The end of the beeline: how pesticides stop bees from finding their way home" *The Guardian*, 30 March 2012, p 3. Alison Benjamin and Brian McCallum, *A World Without Bees*, Guardian Books, 2008.

10. Communities

10

Community orchards

Ever since the environmental charity Common Ground came up with the idea of "Community Orchards" in 1992, more and more people in Sussex have banded together to plant new orchards or restore old ones for community use.

These unique public green spaces are designed to be open and accessible at reasonable times and looked after by local volunteers, who share in the harvest. Sometimes the orchards are owned or leased by the community group or voluntary organisation, or by the parish council or some other tier of local authority. Big or small, in town, city or country, in housing and industrial estates, hospital and school grounds, community orchards are nothing if not "varied and variegated".[1] Their role in people's lives is multifaceted too – as places to "relax, play, work and learn," write Sue Clifford and Angela King of Common Ground. In many places, the Community Orchard is becoming the equivalent of local woodland a century or more ago – a communal asset and a focal point, not least for Apple Day celebrations at fruit-harvest time.[2]

In Sussex, scores of Community Orchards are thriving. The Ringmer Community Orchard, for instance – now planted with 70 fruit trees – was established in December 2005, in a joint project led by Ringmer residents with support from Ringmer Parish Council, Lewes District Council and Common Cause, a Lewes-based community development organisation. The trees are planted in a field bordered by hedges and poplar trees which was made available by the landowner Anthony Tasker at Broyle Place Farm. The aim was to establish a fruit orchard for local use, serving as a model to combat climate change and to be an educational resource for schools. The orchard has more than 30 different varieties of apple, pear and plum – including apple varieties, such as Tinsley Quince, Saltcote Pippin and Alfriston, that have their roots in Sussex towns and villages.

Versatile and verdant: at Ringmer Community Orchard, more than 70 fruit trees are now well established, including this Ribston Pippin – its apples have an intense aromatic flavour, can be used for cooking and are good for making juice and cider too

Training days, open to the public and organised by Common Cause, share expertise on apple pruning and ways of looking after the orchard. Students at Ringmer Community College contribute to orchard tasks, and there are workshops too for local primary schools. Volunteers convene social events, and on summer nights, help naturalists to set up moth lightboxes and bat detectors to identify just some of the wildlife that makes use of the orchard.

The orchard is mown in August to encourage the spread of wild plants in the grassland and by July the grassland under the trees is covered with creeping buttercup. Part of the orchard along a hedge is sown with a wildflower seed mixture to provide a habitat for pollinating insects.

By September the Ribston Pippin, Falstaff and Lord Derby are full of fruit. Orchard members pick the apples, pears and plums with some of the fruit sold in the local shop in Ringmer or given to local schools. The fruit is also made into apple juice and bottled. In January, there is wassailing, reviving an ancient celebration. Future plans for the orchard include the construction of a timber-framed apple store where tools, apple presses and other equipment can also be stored.

Peter May, who has overseen the progress of Ringmer Orchard, and Bryn Thomas, of Brighton Permaculture Trust, have been key to the restoration of two orchards at Stanmer Park near Brighton. With the help of Permaculture Trust volunteers, the South Downs Volunteer Rangers, and others, they have over the years restored both the Home Farm Orchard and the Old Orchard at Stanmer. Brighton Permaculture Trust now holds the lease on both these orchards. The orchard expertise of Peter and Bryn has also been instrumental in a large-scale Sussex-wide project which has supported the creation of community and school orchards across the whole area. Action in rural Sussex and Brighton Permaculture Trust had already collaborated in planting over 20 small orchards in schools when they were awarded funding for an innovative project called Local Fruit Futures. Designed to support sustainable food production, biodiversity and access to healthier foods, the project has set out to make it easier for people to enjoy locally grown fruit, and to promote the growing of that fruit in environmentally friendly green spaces.

Hands on: pupils from Rodmell Primary School help to mulch a newly planted Sussex apple tree in the neighbouring Monks House orchard

To date a thousand apple trees of old Sussex varieties have been propagated by Peter May, Frank P. Matthews Ltd, and Patrick Treherne at Herons Folly Garden (Mayfield) for planting in the new community and school orchards, with surplus trees available for people to plant in their own gardens or orchards. Members of the team have worked with people of all ages in more than forty locations including schools across Sussex as well as housing estates, urban parks, and village greens – showing participants how to plant and look after fruit trees as well as cooking and sharing apple-based food and drinks. At each orchard, a team of people volunteered to look after the young fruit trees, with Bryn and Peter making follow-up visits to check on the trees and assist with aftercare in what one school referred to as their "Apple Ofsted"! The project team has also produced a book for children, *Our Apple Orchards*, as well as this present one, and organises an Apple Day celebration at Stanmer Park each September. At Stanmer, examples of every Sussex

Home Farm Orchard at Stanmer in summer: a perfect picnic spot for pupils from New Horizons School in Hastings, researching orchards in preparation for planting their own fruit trees

Plum harvest at Grays Infant School in Newhaven

apple variety have been planted in the recently extended Home Farm orchard, and their progress is carefully monitored. Funding has allowed a wheelchair accessible path to run through the orchard – school and community group visits are welcomed in all seasons. Advice is offered on the restoration of old orchards and the design of new ones for community use and there is an ongoing programme of training days in orchard skills (see the last section of this book for more details).

The small orchards planted under the auspices of the Local Fruit Futures project in East Sussex include Claverham Community College at Battle, Churchwood School at St Leonards-on-Sea, Southdown Junior School, Newhaven, West St Leonard's Primary School, Langney Primary School in Eastbourne, Chyngton Primary School in Seaford, Grays Infant School, Newhaven, Uckfield Community Technology College, Willingdon Community School, St Mary's School, Maynards Green, the Crouch Gardens, Seaford and Burwash Village Sports Ground.

Orchards in the Brighton and Hove area include those at: Whitehawk Primary School, Moulsecoomb Primary School, Patcham Special School, The Cedar Centre, St John the Baptist Primary School, Dorothy Stringer School, Varndean School, Saunders Park, and St. Luke's Church all in Brighton; and Hove Park School, Rudyard Kipling Primary School at Woodingdean, Mile Oak Primary School at Portslade, and Wish Park in Hove. Demonstrating that fruit trees can be fitted into really quite restricted urban spaces are the small orchards on cordons planted at the Phoenix Community Centre and at London Road Railway Station, Brighton. The first trees at Craven Vale estate were also planted through the project.

Generations united: a Burwash cub scout and leader Emanuel Flecken (aka Red Fox) put the finishing touches to a newly planted tree

Orchards in West Sussex include: Billingshurst, planted by primary school pupils as the first stage in a community food and flower growing area established by Billingshurst Friends of the Garden; Rotherlea Care Home with pupils from Petworth Primary School; The Globe School, Lancing; Fittleworth Primary School; Hollycombe Primary School in Milland; Glebe Primary School, Southwick; a green adjoining Kirdford Village Shop; St Margaret's Primary School at Ifield, Crawley; Burgess Hill Link School, Imberhorne Secondary School, East Grinstead; St Wilfrids Secondary School, Crawley; and Manor Field Primary School at Burgess Hill.

Orchard plantings are accompanied by an apple cookery session often led by Community Chef Robin van Creveld. Sometimes planters and cooks get together to share a celebratory apple-themed meal.

One school, Manor Primary in Uckfield, celebrated its new orchard by organising a whole day of apple-themed science, art, poetry and maths activities alongside the planting and cookery.

Apple chef: Community Chef Robin van Creveld cooks in the January sunshine, using apples in imaginative ways, for all who worked up an appetite helping out with the orchard planting at Wish Park, Hove

As already mentioned, in Brighton, the Community Association at Craven Vale housing estate planted eight trees with help from Local Fruit Futures, and are now planting many more, with further support from Brighton Permaculture Trust and funding from The Big Tree Plant and Harvest–Brighton & Hove. Forty-eight fruit trees will ultimately be planted, and there are plans to establish wildflowers around the trees in the spring with the help of the Sussex Wildlife Trust. Grant Scott of Craven Vale has been the powerhouse behind the ambitious orchard-in-the-making, which will utterly transform the once-bare and windswept landscape around the housing estate. As he explained, "We wanted to regenerate a public place and get the community involved with planting and looking after rare Sussex varieties of apple trees." Grant has inspired a groundswell of support among local residents, including an active youth group and an arts group.

Brighton Permaculture Trust and Action in rural Sussex continue to support small orchard planting and orchard skills training, and are spearheading the planting of community orchards on a larger scale in rural and urban settings. They have already begun on the next stage of making orchard fruits the enduring, year-round element in a sustainable food system. As the trees increase their bounty year on year, consumers of fresh fruit, however keen, can't keep up with the supply, so the excess fruits need to be preserved or transformed into juice, jams, chutneys, soups, cakes and cider. Local apple presses, community cookery kitchens and sale through local outlets are all now operating and are ripe for expansion.

Other community groups are also fuelling the orchard revival across Sussex.

In Brighton, Elspeth Broady has been the driving force behind a new small orchard planted at London Road Station. The orchard has cordons of four old Sussex varieties – Crawley Beauty, Tinsley Quince, Saltcote Pippin, Mannington's Pearmain – as well as two pears and a plum tree.

Prime mover: Grant Scott, chair of Craven Vale Community Association (second left), admires a new addition to the burgeoning community orchard, joined by local pupils, residents and orchard experts Stephan Gehrels (first left) and Bryn Thomas (fourth left)

She tells the story: "The orchard idea came from a chance meeting with Joanna Harper from Action in rural Sussex. She was passing the garden one afternoon when we were working there, and suggested we could plant an orchard. I think I was rather taken with the challenge of growing fruit trees in such a small space. Our little edible-growing plot is dominated by a large south-facing wall, and we'd talked about trying to grow something up it that would have impact for passers-by. Fruit trees were really intriguing: to show this was possible in a small urban setting (a railway station), and to grow local varieties – this seemed like something that would interest local people. And the idea of growing something that could last for 50 years or so was also intriguing – up until then, we'd been thinking about annual vegetables. It was a real challenge constructing the raised beds, getting the soil and compost and designing/building the growing wires, but the fascination of local people with the orchard has made it all worthwhile. So many people stop and ask about it now. The nicest comment was one I overheard: one neighbour I didn't know to another neighbour saying, 'Isn't our orchard looking great?' It seems that the orchard has been adopted by the local community. Two neighbours told me that they have brought visiting friends up to see it."

Orchards come in all sizes: London Road Station in Brighton now hosts a thriving small-scale community orchard of cordon apples, thanks to a chance meeting between local resident Elspeth Broady and a member of staff from Action in rural Sussex

Mark Allery, Commons Manager of the Lynchmere Society, and an accomplished pole-lathe turner and woodsman, described another, quite different, new orchard:

> "Up here on the north-west frontier of West Sussex in Lynchmere we have established a small community orchard along the edge of a field. We have planted about thirty trees of half standard size, all traditional apple varieties, many from Sussex and the surrounding counties, as well as a crab apple, Victoria plum, a quince and a walnut."

A delicious Sussex variety, Saltcote Pippin, originally from near Rye is now doing well at London Road Station

The Community Orchard opened in June 2011, supported by a grant from the South Downs National Park Sustainable Communities Fund. A new Society notice board lists tree varieties within the orchard. Mark had also helped to establish two small orchards planted by the Blackdown National Trust team.

In November 2011 Chichester Community Orchard of 30 apple, pear and plum trees was planted next to the Festival Theatre in the county town – the brainchild of Transition Chichester, who will manage it with Chichester District Council. The trees were planted by volunteers including children from local schools, members of Transition Chichester and local Rotary Club members who provided the money for the trees. Chichester District Council and the Open Spaces team also took part. The community orchard will provide fruit for local residents and visitors to the park and more trees will be added at a later date. The orchard was designed in conjunction with Chichester Festival Theatre with advice from Local Fruit Futures. The trees are laid out in a gentle arc around the theatre buildings with varieties such as Bramley's Seedling, Charles Ross, Pixie and Egremont Russet. A large wildflower meadow is being sown under the orchard trees.

Dramatic developments: a new community orchard has been planted next to Chichester Festival Theatre

Local resident of East Chiltington, Suzie Hopkins offers this account of her village's new orchard:

> "The East Chiltington Trust was formed following the purchase of a field in the middle of the parish to save it from development. The land was bought with funds from the parish council and donations from local residents. The Trust consulted with local residents about the preferred use of the field. This consultation revealed a wish to retain the field in agricultural use with associated environmental projects. We therefore decided to let the majority of the field to a local farmer, retaining half an acre in front of the pub for the planting of a community orchard. Funds for the orchard were raised by local residents and through local sponsorship. We also received grants from Lewes District Council and Grass Roots. Residents were consulted about the positioning of the orchard and the types of fruit to be planted. A variety of those suggested were ordered for planting including, apples, pears, plums, quince and a medlar.
>
> Twenty-three standard trees were planted on 13th December 2010 which will grow to create a traditional orchard. We had excellent weather and an abundance of volunteers who were helped on the day by Peter May, a local orchard expert. Since then we have held work mornings to prune and care for the trees, plant bulbs and locally sourced wildflower seeds with advice from Anthea Stevens. This project has been highly

> successful, both in terms of a healthy young orchard, and in creating a focal point for our small rural community".

At Selsey, a Community Orchard is being planted in Manor Green Park. All the trees are Sussex apple varieties. Under the auspices of the Selsey Town Council, local tree wardens oversee the scheme. The idea is that each tree is "adopted" by a local resident or a community group – children from First Steps Nursery planted theirs last year. The tree wardens then provide each adoptee with a detailed information pack about the apple tree, its fruit and its aftercare. Wardens keep a permanent record of all the sponsored trees and their adoptees, who can also provide a brief citation to go with the tree's official record in the book.

At Bognor, members of the Greener Bognor Network are creating the Rife Community Orchard, with Edible Hedging, on a small strip of abandoned land, south of a stream, Aldingbourne Rife. Overgrown with scrub, local people reckon the site had not been touched for at least a decade. Initially, the group plan to plant three apple trees and a plum tree, and a hedge of crab apple, elder and dog rose, using trees offered by the Woodland Trust in its "wild harvest" scheme. They hope to go on to plant damson, bullace, mirabelle, cherry plum and sweet chestnut.

In early December 2011, a new Community Orchard was planted at Cortis Avenue Wildlife Garden in Worthing. Project co-ordinator Hille Cook explains:

> "The orchard and garden is on land owned by Worthing Homes Ltd and disused for years after serious levels of fly-tipping. Their support for the project enabled local residents to work together in a community action project to transform a wasted piece of land into a flourishing green space in the heart of Worthing."

Six old Sussex varieties were planted: First and Last, Crawley Reinette, Bossom, Golden Bounty, Ducks Bill and Saltcote Pippin. The trees are on vigorous rootstocks (MM111 and M25) which will result in large trees that are more resilient in drier, poorer soils. The trees all come from the Brighton Permaculture Trust's nursery at Stanmer, apart from First and Last, which was supplied by Keepers Nursery, East Farleigh, Kent. The new orchard was made possible thanks to donations from staff at the Environment Agency, with support from Brighton Permaculture Trust and Worthing Homes, and by members of Transition Town Worthing. Hille Cook adds: "Our new community orchard goes some way to creating a new long-lived and distinctive food heritage for the town, with an orchard that can produce food for up to 100 years." The group has plans to invite Sompting Village Morris for apple-themed celebrations in the future. The Morris band already conduct an annual "wassail" or "apple howling" in Tarring village in Worthing, preceded by a torchlit procession.

Also in December 2011, a new Community Orchard was born in Littlehampton. Twenty pupils from the town's River Beach Primary School joined fifteen volunteers from the community to plant 27 fruit trees, including Sussex varieties, in a field at the rear of Maltravers Park. Staff in Arun District Council's parks department actively supported the project, which is seen as a pilot scheme for many more orchards in the district. The council has launched the scheme with Brighton Permaculture Trust, the Arun Wellbeing Team, and Arun and Worthing Mind volunteers from the St Winifrid's Community Gardeners, who will ensure that the orchard will be well looked-after. "We are now hoping to establish a new "Orchard Group" – a committee of volunteers to help decide on the future management and events associated with the orchard and Maltravers Park," says Helen Wilson, Arun's senior community parks officer, including apple pressing, fruit cookery les sons and annual Apple Day celebrations.

At Staplecross, Ewhurst Green Community Orchard runs its own Apple Day celebration, juicing garden apples with a press, and a school orchard has been planted with pupils.

At Hassocks, a new Community Orchard of apple and plum trees was planted in February 2012, in land behind the Pauline Thaw Age Concern Day Centre at the instigation of Hassocks, Hurstpierpoint, Keymer and Ditchling Transition Group.

Community orchards are not always entirely new creations – some focus on the restoration or extension of an existing orchard. Between Lindfield and Haywards Heath, at the 15-acre Scrase Valley Nature Reserve, a community group has set about rescuing neglected, overgrown orchard trees, some of which are impressively large. The old orchard forms part of an important patch of green, including woodland, scrub, grasslands and marsh. New trees will also be planted at the new Community Orchard on the reserve.

Growing well: the standard trees Randolph's Orchard at Iden are teenagers now and already a good height, allowing sheep to graze the rich grassland beneath them

The villagers of Iden and the Crowhurst Community Agriculture project are helping to manage Randolph's Orchard. Over 200 apple, pear and plum trees of 40 different varieties have been planted by Christopher Strangeways since 1998 on a sloping site to the south of Iden. There are large standard Rosemary Russet, Golden Noble and Blenheim Orange while sheep graze underneath the trees. During the autumn fruit is picked and apple juice pressed and through the winter months there are training courses organised by Plumpton College on orchard planting, management and pruning.

At West Grinstead – alongside a former pub which became a Little Chef and is now an independent restaurant serving local produce – a new Community Orchard has been established in an old derelict orchard. The initiative, brainchild of the West Sussex Rangers' Service, was funded by a Countryside Stewardship grant, with the-then Little Chef providing free parking for educational visits. In 2010, the orchard hosted the Horsham and District Apple Day 2010. The current restaurant, named The Orchard, makes use of perhaps ten different apple varieties from the orchard in its menus.

Apple, drawn by Rhys, Castledown Primary School, Hastings

Children from Wisborough Green Primary School helped to plant a new Community Orchard at Fishers Farm Park, near Wisborough Green, planted with Sussex apple varieties. The new orchard was planted under the auspices of the West Weald Landscape Project, organised by Petra Billings and based at the Sussex Wildlife Trust, which is encouraging landowners to value old orchards for their contribution to the "connectivity" of the landscape via corridors of wildlife-friendly habitats.

And on National Trust land at Slindon, an old orchard has been extended for the benefit of the villagers, and as part of the village's Diamond Jubilee celebrations. Mike Imms reports:

> "Our orchard is planted! Around 25 trees were planted in January 2012 on a section of field adjoining the village hall, and bounded by a community field and Slindon Forge, where we are creating a community shop, cafe and info hub. So the orchard is at the heart of the village and connects directly with other community assets.
>
> The funding for the orchard came from villagers who 'sponsored a tree' – and many of them, and others, came along to the communal planting session. The National Trust owns the freehold to the orchard (and the community field) which is licensed to the village hall committee, and has also done a huge amount of work helping with machinery, expertise – and labour. Katie Archer of the NT is a key member of the orchard committee."

Blossom, drawn by Byron, Castledown Primary School, Hastings

In Sussex, many community orchards are supporting the revival of ancient celebrations, including that of "wassailing" – from the Anglo-Saxon *waes haeil* – to be healthy.[3] Often known as "howling" in Sussex, participants sing to the apple trees on Twelfth Night, beat them with sticks and pour ale or cider on their roots to ensure a good crop in the autumn. Howling's origins are genuinely old – unlike some other "traditional" customs, which in fact date back only to Victorian times – with documentary evidence surviving from the seventeenth century at least. Giles Moore, rector of Horsted Keynes, wrote in his diary on 5 January 1656, "Gave to the howling boys sixpence." In recent years, Sussex Morris Men have revived the howling custom, and in 2010, the Chanctonbury company visited Old Mill Farm, Bolney, placing a spiced and cider-soaked wassail cake in a bough of the oldest tree – the Apple Tree Man, the guardian of the orchard. After the ceremony and "hullabaloo", the celebrating company sampled cider made on the farm.[4]

Ancient celebrations revisited: wassailing at Old Mill Farm, Bolney

Finding homes for unwanted apples

Brighton Permaculture Trust is also harvesting unused fruit in the Brighton and Hove conurbation. This is the Scrumping Project, part of Harvest–Brighton & Hove, a series of initiatives that aim to help more people in Brighton and Hove grow their own and eat local food, by sharing skills and making more land available for food growing. The project collected three tonnes of fruit in 2010 from trees in the city's back gardens and open spaces. The fruit was juiced at community events, shared among volunteers, baked into cakes and crumbles, or converted into chutneys and bottled juices.

The Fruit Tree Project in Sussex and Kent, launched by Chris Blastock and Mark Elliston, is a community project, based in Hastings and St Leonards, to prevent surplus or unwanted local fruit going to waste. Residents can register a fruit tree and arrange for volunteers to pick the fruit – leaving at least 25 per cent with the householders, if they would like it. The not-for-profit organisation is also hoping to find and run unused orchards, including harvesting and running pruning and grafting workshops alongside planting new orchards for the future. The group use the fruit in a variety of ways, selling fresh fruit, juice and preserves from market stalls, raising funds for the project as well.

Willing helpers wait their turn on a traditional press overseen by Stephan Gehrels, who, when not leading school Eco Clubs, enjoys goading children and adults alike into making their own apple juice, here at Apple Day at Stanmer

In Lewes, a project called OctoberFeast is working to map the town's apple trees and help householders to harvest unwanted fruit, so that it can be turned into juice or distributed round the town. The project hopes to discover old varieties and encourage the planting of new trees.

Orcharding legacies

The benefits of community orchards in all their variety are impossible to catalogue definitively, but may yet prove to have their strongest impact on younger generations. Perhaps apple orchards will boost the quality of life most profoundly by helping to combat a "nature-deficit disorder" among modern-day children, who are placed in "what amounts to well-meaning, protective house arrest, their play reduced to the manipulation of joysticks and keyboards.[5]" Natural England, in a survey in 2009, found that fewer than 10 per cent of children play in woodlands, countryside and heathlands, compared with 40 per cent of adults when they were young. Community orchards can reconnect children with fresh, natural food, as the consumption of fruit and vegetables continues to decline. In 2010, a UK "fruit and vegetable task force", convened to look at ways of increasing the production and consumption of fresh produce, advocated an increase in the provision of community growing spaces – including allotments, community gardens and community orchards. These special places provide "green exercise" and "outdoor living rooms" – perhaps helping to fend off a range of childhood ills including obesity, attention-deficit disorder, isolation and depression.

A plenitude of "Scrumped" chutneys, pickles, jams and juices, all made from fruit picked in Brighton and Hove which would otherwise have gone to waste

School orchards planted by Brighton Permaculture Trust and Action in rural Sussex together with other school orchards now number more than seventy – over ten per cent of schools – and there may well be over forty community orchards in Sussex. Some of these new orchards have been helped along by the same two organisations, and others by Transition Town groups, by Lewes-based Common Cause, by Parish Councils; others have simply come about as a result of like-minded people coming together. The trend seems set to continue and the network of apple-growing enthusiasts in Sussex continues to widen.

All welcome: the orchards at Stanmer offer visitors to Apple Day a chance to enjoy a sense of place and peacefulness

Endnotes for Chapter 10

1 Common Ground, *The Apple Source Book*, London: Hodder & Stoughton, 2007, 49.

2 Common Ground, *Community Orchards Handbook*, Totnes: Green Books, revised edition 2011.

3 S. Clifford and A. King, *England in Particular*, London: Hodder and Stoughton, 2006, 430.

4 J. Simpson, *Folklore of Sussex* Sroud: Tempus 2002, 99-102.

5 R. Louv, *Last child in the woods: saving our children from nature-deficit disorder*, London: Atlantic Books, 2010, and see his article 'A timely truth' in *National Trust Magazine* Spring 2011, 34-7.

II. Once & Future

11

Once and future orchards: sharing with nature

Astonishingly, almost a thousand old orchards can still be found in Sussex today. These stalwart survivors have been "re-discovered" only recently, as part of a nation-wide hunt for traditional orchards. The survey, organised by the wildlife charity the People's Trust for Endangered Species (PTES), used detailed aerial photographs to pinpoint clusters of old apple trees. For purposes of the survey, traditional orchards have been – defined as groups of at least five well-spaced and good-sized trees, planted on vigorous or semi-vigorous rootstock, set in permanent grassland which is managed in a low-intensity way. Under those criteria, East Sussex turns out to have 253 traditional orchards, while West Sussex boasts 668, covering a total of more than 300 hectares in all.[1]

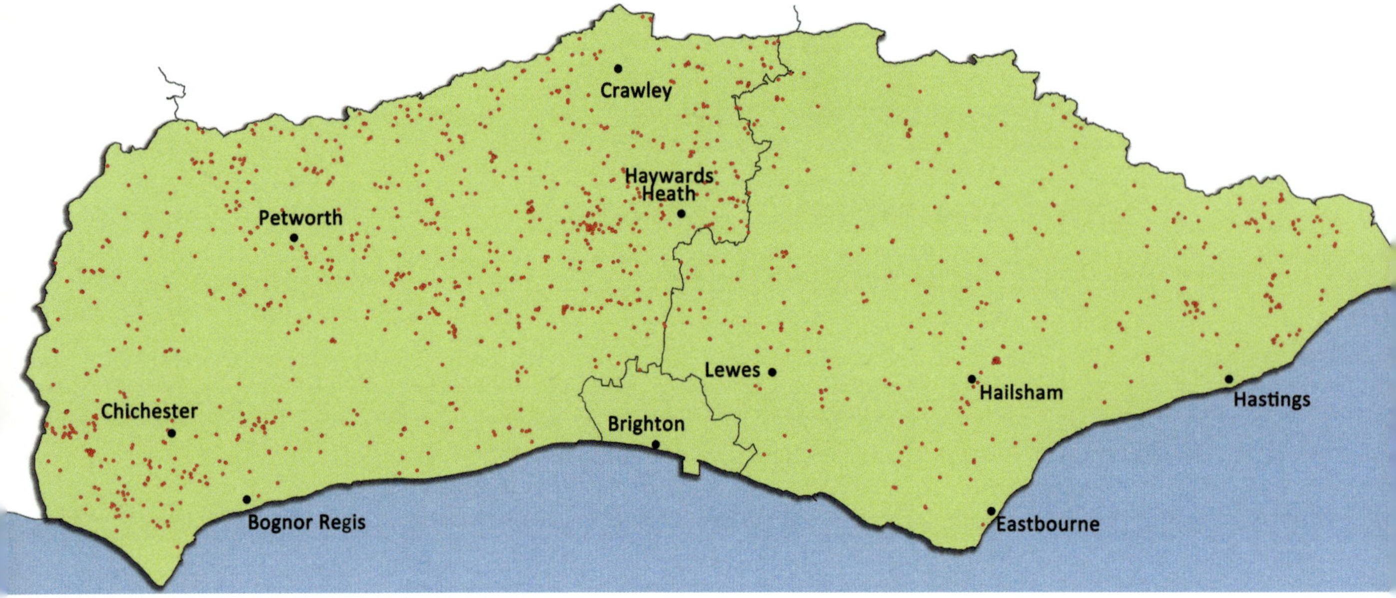

Fruitful inheritance: traditional orchards in Sussex, mapped by the PTES survey, number in the hundreds. Although many are small and in need of gentle restoration, they testify to Sussex's enduring orcharding culture

Not many people have had the pleasure of spending time in one of these places – most survive in country gardens or tucked away at the corners of agricultural land. But old orchards are welcoming places, typically graced with tall, mature trees planted with plenty of space between them, and often with rich grassy, flower-rich swards underfoot. They demonstrate better than anything else that wildness and humanity, nature and culture, can work together, to mutual benefit. Survivors from a world we have almost lost, they may hold lessons for the future.

Today's orcharding renaissance began with Common Ground's Save Our Orchards campaign, launched in 1988.[2] Just in time, this unique environmental charity has sparked a profound revival in our nation's rich orcharding legacy. Its great achievement has been to show why saving old orchards, as well as planting new ones, is so important. Orchards are dynamic habitats, changing in character as the trees age, and growing in biodiversity as all the plants' roots, and the living soil communities they support, remain undisturbed year after year. And it's not just the natural world that benefits. These beautiful and productive places connect us to localities in ways that matter to us – each with its own history and idiosyncrasies, shaped by time and happenstance, geology and the weather, they reflect and foster local distinctiveness. "The loss of old orchards is a loss to landscape, wild life, culture, employment and local meaning," writes Common Ground's co-director Sue Clifford.[3] The loss of our orchards has also eroded our previously taken-for-granted reliance, in part at least, on local supplies of fruit – whether from professional growers or from smaller orchards in gardens and farmsteads.

Old orchards are biodiversity hotspots

Light filtering through apple trees encourages an understorey rich in plantlife, not least cow parsley

As Common Ground's rescue campaign gathered momentum, the official world of wildlife conservation was slow at first to join in. To the nature professionals, these "man-made" places with "domesticated" trees didn't seem to qualify as proper wildlife habitats. But in 2007, at last, traditional orchards were designated as a "priority habitat" under the UK Biodiversity Action Plan – ranked alongside our finest places for wildlife.[4] While this accolade gives no formal legal protection to orchards, it does add weight to campaigners fighting to save orchards targeted by planning applications, and should help orchard owners to receive funds from Defra's "environmental stewardship" scheme. At the moment, some 2,800 hectares of traditional orchards on farmland are being managed within this scheme in England as a whole. Some counties have included orchards in Local Biodiversity Action Plans, though not yet East or West Sussex, or the city of Brighton and Hove. And it is possible to petition a local authority for a tree protection order on one or more old fruit trees that are no longer part of commercial fruit farms and which are visible from roads and footpaths.[5]

Why are old orchards so good for nature? Nature conservationists have struggled to define "the orchard habitat" in the way they describe chalk downland or heather moorland, for instance, simply because no particular cluster of animals or plants are unique to orchards. Yet these places are special because they are so various, so hard to pin down. Blessed with a hotchpotch of microhabitats, old orchards offer a mosaic of niches, essential for a great many organisms that need a variety of habitats cheek by jowl to accommodate the radically different stages of their lifecycle – from egg to caterpillar to chrysalis to butterfly, for instance. An orchard can be simultaneously woodland, meadow, pasture and hedgerow, and might even have a pond or an old stone wall or two. Orchards are by their very nature hybrid entities, neither quite field nor woodland, but something in between, and astonishingly diverse in size, shape and make-up, with a character and atmosphere that varies with locality.

Built to last: an apple orchard at Mottynsden Farm, Burwash, in the High Weald with permanent pasture grazed by sheep

They most resemble wood-pasture, another "man-made" habitat now recognised as among our oldest natural landscapes. This open landscape – now

Wild fruit trees may have evolved amidst small copses and open glades not so very different from modern parkland – glimpsed here at West Dean, where the walled orchard is well-stocked with old Sussex apple varieties

Facing page: Fruitful symbiosis: lichens develop on the bark as apple trees age, and do no harm to the trees. Often a good sign of relatively unpolluted air, these algae-fungal partnerships provide shelter for beneficial insects too

associated with National Trust country estates, and characterised by large, well-spaced trees with grassland grazed by livestock – was once created by the grazing and browsing of a host of wild mammals including deer and aurochs. The primeval wildwood, typically celebrated as a great closed forest of towering oaks, might actually have looked much more like today's parklands, with "natural orchards" of wild crabs, sloes, bullace plums and cherries.[6] Fruitful landscapes have always been part of our history.

An intensive study of three traditional orchards in the west country some years ago revealed a total of 1,800 species making use of these places.[7] A second survey of old tall-tree orchards in the region, carried out by Natural England in 2009, produced similar results: orchards that are managed sympathetically, without dousings of pesticides, can support a remarkable diversity of fungi, rare invertebrates and plants.[8] Old orchards in Sussex have yet to receive the close attention they deserve; the staff at the Sussex Biodiversity Record Centre at Woods Mill hope that expert naturalists will begin to target these places, to find out what's living there and fill the glaring gap in its records. But botanists have kept an eye on the plantlife in Old Stanmer Orchard, near Brighton, and in among the standard apple trees, meadow cranesbill, butterbur and dog's mercury thrive with the occasional bluebell, enchanter's nightshade and twayblade.

On the shaded southern flint wall, harts tongue fern and maidenhair spleenwort have found a home. Lichen experts recently surveyed the two Stanmer orchards, and found a total of 36 different sorts of lichen, including two of the three species most sought after by long-tailed tits. These beautiful little birds build exquisite spherical nests which they painstakingly cover with some 3,000 flakes of these grey-green-and-white lichens, hooking each flake onto the nest using tiny filaments on the lichen's underside like Velcro. By reflecting light, this mantle of lichen helps to conceal the nest – and the birds need to find just the right species of lichens to do the job. As the Stanmer apple trees continue to mature, the wealth of wildlife they support should grow richer still.

Above: Venerable apple trees can go on and on, however much they lean or even rest their aged limbs upon the ground, like this Bramley's Seedling in the Old Orchard at Stanmer

James Marsden, a conservationist with his own old orchard in Herefordshire, has estimated that the number of wild species in an orchard roughly doubles in orchards with apple trees that are at least 25 years old – and an apple tree can live to be 120 or more.[9] By contrast, today's commercial growers typically plant small, densely packed, bush-like rows of apple trees, resembling vineyards, that bear fruit quickly. When the productivity of these trees begins to decline, after they have grown for 15 years or so, they are pulled out and replaced, disrupting the land and its living communities. Old orchards are havens for a host of interesting and sometimes unusual organisms that need continuity of habitat over time.

The tall, aged trees themselves provide refuges for all manner of life forms. As the trees develop cracks in their bark, crevices abound for all sorts of beneficial insects such as ladybirds and spiders, as well as nooks and crannies for hole-nesting birds and bats. Older branches and trunks develop green mantles of lichens, mosses and liverworts which provide food, refuge and camouflage for a profusion of beetles, bugs, spiders, ants and moths that stoke the

food webs on which birds and mammals (including humans) depend. And when the trees at last start to die, they offer food and shelter to a fascinating and rare assemblage of deadwood specialists, not least the large and impressive beetle known as the noble chafer, which finds a toe-hold in just six English counties, almost exclusively in old orchards.[10]

This flagship beetle has yet to be found in Sussex orchards, although a population has recently been re-discovered in Kent, and there are historic records in West Sussex. But the noble chafer's greatest significance lies in its role as a signpost to a vast community of organisms that thrive in decaying apple wood, a web of life which may embrace several hundred species of beetles, flies, bees, wasps, spiders, molluscs, centipedes, millipedes, false scorpions and more, according to Roger Key, an ecologist at Natural England.[11] Living in old apple wood, this wealth of invertebrate life in turn nourishes frogs and toads, voles and moles, foxes and badgers too. Blackbirds and blue tits, treecreepers, flycatchers, thrushes and finches join in the feast and may nest in tangled branches. Bird species regarded as farmland specialists, woodland generalists and woodland specialists can all make good use of old orchards.

Long-tailed tits visit old apple trees to harvest flakes of lichens with which they carefully camouflage their nests

Even the ground in which the apple trees spread their roots is special. When undisturbed for decades, its soil communities are rich in earthworms and symbiotic fungi, with species-rich grasslands sporting orchids or a host of other wild flowers, as well as the delicate waxcap fungi. Holes in old apple trees can house several species of bats, little owls and even Britain's smallest and most elusive woodpecker, the lesser spotted, no bigger than a sparrow and seen regularly in Stanmer Old Orchard until two decades ago. Mistletoe has a particular affinity to the apple tree, as does the mistletoe marble moth – speckled with white, fawn and cobalt-blue, to mimic a bird dropping.[12]

Denizen of deadwood: this rare and beautiful beetle thrives in old stumps and decaying wood found on old apple trees

Old trees offer sustenance for fungi too, not least chicken-of-the-woods and the weeping bracket, as well as the very rare "orchard tooth fungus" which smells of pineapple when fresh – all feed on the dead heartwood of old apple trees. "These fungi do no harm to the trees and the hollowing they cause may actually benefit the tree by recycling nutrients and improving its ability to withstand strong winds," Natural England explains in its helpful technical information notes on traditional orchards.[13]

The apple blossom provides a banquet of pollen and nectar for a host of pollinating insects that perform the vital role of carrying pollen from flower to flower. Most apple cultivars are "self-incompatible" and won't set fruit unless their flowers receive a visit from a friendly insect laden with pollen from another variety of apple tree of the same species. Honeybees, wild solitary bees, and bumblebees are prime pollinators, and some apple varieties have flowers that are best pollinated by the bigger bumblebees.[14]

Caring for orchards: examples from Sussex

No wonder, then, that old orchards are now being valued for their contribution to the network of old hedgerows and parkland and patches of woodland that ideally form corridors of green spaces. The West Weald Landscape Project, run by Petra Billings at the Sussex Wildlife Trust, is championing traditional orchards for this very reason, and advising landowners on how to look after their prized old fruit trees as well as encouraging them to plant new orchards.[15]

Decades-old orchards abound particularly in the west of the county, not least at Penny Aschan's land on Bedham Lane, Fittleworth, where 30 trees planted in about 1970 are still producing a glut of the "most beautiful apples" for which Penny struggles to find homes. Nearby at Little Bognor House, head gardener Sean Bowyer looks after a 26-acre orchard, first planted in the 1920s by General Sir Ivor Maxse (mentioned in chapter 7) and predominantly Cox on M9 dwarfing rootstock, producing trees about 8 feet (2.4m) tall. There are a few bigger trees, and some are probably octogenarians. "There is no concerted effort to pick the fruit because we have no market for it," Sean explains, so most fruit falls to the ground each autumn, where it feeds birds and mammals galore. "A flock of about 40 fieldfares take up residence in the orchard every winter," says Sean, "and I am certain that there is considerably more wildlife present at any time in the orchard than there is in surrounding arable fields." The orchard with its grassy sward is only lightly managed, without chemical sprays.

At West Burton near Pulborough, Jean and Steve Jackman tend an award-winning wildlife-friendly garden with two "oldish" apple trees they inherited, both bedecked with mistletoe. Until about eight years ago, they enjoyed seeing

Orchard enthusiasts: vole, hedgehog, collared dove, green woodpecker and blue-tit all visit orchards, and blue-tits in particular may even help to keep apple pests such as codling moth at bay

Insects such as wild bees and honeybees act as vital pollinators, while a whole host of other creatures enjoy feasting on windfalls, including fieldfare in winter, wasps and butterflies like the Red Admiral above

mistle thrushes eating the mistletoe berries, "but there are none to be seen now sadly". One tree, probably 60 years old, is a Bramley with "superb fruit," says Jean. "The apples that fall we leave on the ground for blackbirds and in a bad winter also see fieldfares on them, digging them from under the snow." Deer eat codlins off the ground, and the pears, "small fruits, very sweet," are much loved by the wasps.

In Wadhurst, one privately owned orchard has become a community asset. Shirley Rothera started planting apple trees in the 1970s and now has a Wealden orchard stretching over a couple of acres, stocked with 25 different varieties, including a Wadhurst Pippin. She took her inspiration from the five fruit trees she inherited when she moved into her cottage 50 years ago: two pears, a Victoria plum and two apples, including a Bramley. She's now got a walnut tree too, though she has to compete with the grey squirrels as the nuts ripen. She makes good use of her harvest, pressing her own apples and freezing the juice, and distributing the fruit through the Wealden Co-operative in Wadhurst. "Apples are a wonderful food, and there's a nucleus of local people who help pick and prune and make juice and share in the harvest," Shirley explains. She doesn't spray her trees, and enjoys the wildlife that make use of her orchard (apart from the squirrels!).

Older still is Jo Simmons' traditional orchard in the middle of her 13-acre smallholding at Fulking, at the foot of the South Downs. She knows who planted it and when – her extensive orchard of some 90 big trees and a damson hedge was planted in 1911 by John Tamkin. Some of the fruit trees are "massively tall," she says, with "apples like footballs." Although some trees have died and collapsed, most are alive and productive even though they have lots of deadwood – "the woodpeckers love it here, they nest every year in the tree holes, and the noise they make is incredible – to me it heralds the start of spring when we hear the woodpeckers hammering." When she and her husband Trevor took over the orchard in 1996, "you couldn't see the fruit trees in all the bramble and elderberry – it has taken us years to clear it all." But the trees have survived, Jo suspects, "because it was forgotten, and left fallow all those years".

The Simmons are gradually restoring the orchard, "getting the balance right between keeping the place tidy and manageable and keeping its character". They are taking out dead trees and stumps and replanting, "because I want the orchard to be there in another hundred years." She explains: "We want to get it back up working, which will give it a darn sight better chance of surviving through time." They manage the grassy sward beneath the trees by grazing it with sheep during spring and autumn and mowing it during the summer months.

No one has yet had a go at identifying the fruit varieties, but the orchard has apples, plums and a few pear trees, which might be old perry pears or cooking pears – "some are 40 feet high, it's almost impossible to get at the fruit." If the orchard should turn out to have "weird and wonderful varieties" the Simmons may try to plant new trees grafted from the old stock. The plums are definitely worth saving: they're "sheer nectar", she says. She and her husband are also hoping to put in a cherry orchard – "we want to put our stamp on the orchard – it's our hobby and our passion." They've "made cider really badly" but would be keen to try again with some expert help, and are trying to find someone to manage the harvesting and marketing of all the produce. "We're not particularly interested in making money, but would want to see the fruit put to good use. It breaks my heart whenever I go round Tescos and see not a single English apple," Jo says. "The time is ripe to get this place back up working again."

Centenarians: planted in 1911, this old orchard at Fulking is now being restored by Jo and Trevor Simmons

It is still possible in Sussex to buy fruit direct from a traditional orchard – from, for instance, Tony Eales' Tendrings Fruit Farm at Magham Down near Hailsham – which has been growing apples for a hundred years. A few trees planted in 1910 are still producing fruit, with most trees in their 30s and 40s. Tony's orchard is on the edge of Pevensey Levels, which he visits regularly to record its rich wild life, including his favourite bird, the beautiful and elegant hobby, which hunt dragonflies over the marsh. His old orchard, stretching all around his house, supports a great variety of birds too, especially now that he no longer sprays the trees with insecticides.

On a snowy winter's day, Tony says, it's not unusual to see the orchard full of fieldfares and redwings, flying in to feast on apples he puts out for them. Greater spotted woodpeckers nest in the old fruit trees, and green woodpeckers, blackbirds and thrushes and nuthatches use the orchard all year round. Alongside robins and dunnocks, there are blue tits, great tits, coal tits and long-tailed tits, and bullfinches, chaffinches and greenfinches.

One of many Bramley trees still bearing fruit after one hundred years at Tendrings Fruit Farm

Crows and magpies are in residence, and "jays like the orchard too, and eat the apples, the beggars," he chuckles. Sparrowhawks swoop in to catch their supper, as dragonflies patrol the farm ponds and gardens. "Basil" the buzzard rests on trees at the back of Tony's house, while jackdaws nest in a disused chimney. Tawny owls roost out in the orchard. Dormice wander into the apple store, and badgers are "all over" during the apple season. Butterflies and bumblebees, wasps and hornets, all appreciate the bounty: "I don't mind them," says Tony, though he says he often gets a wasp sting or two, picking the plums. He worries about modern agriculture, "fields full of cereals with no weeds, so there are no seeds or insects for the birds." The starlings are nothing like they used to be, he says, "they don't breed with us now," and nor do house sparrows or house martins or swallows – "we've lost the lot. I haven't had a swallow for ten years." As small-scale, mixed farming – with low-intensity orchards, fields and livestock intermingled – has disappeared, so the ability of our countryside to support a rich living world has dramatically declined.

Good harvest: old-style orchards can arguably be just as productive as modern, bush-style plantations

Recruiting natural allies in the orchard

Old orchards remind us how easy apples are to grow – in ways that augment rather than damage the ecological health of their surroundings. It takes less than half an hour to plant a tree that could be here for the next 80 years and more, producing fruit year after year – apple trees are one of our most reliable sources of food. No wonder the word "windfall" means not only a piece of fruit blown down by the wind but a piece of unexpected good fortune.

In a more sustainable future, one far less reliant on a battery of toxic synthetic chemicals and fossil-fuel fertilisers, fruit-growing on a domestic scale and in public parks and gardens could provide a good supply of nutritious food virtually for free. Apples grown in small orchards – with just a few well-spaced trees of hardier varieties chosen for their resistance to the fungal diseases of scab, canker and mildew – are likely to be intrinsically healthier and less prone to outbreaks of disease and insect pests. Modern intensive agriculture, with its large monoculture plantations, provides economies of scale, but also creates ideal breeding grounds for pests.[16]

Older techniques of good husbandry were labour intensive but surprisingly effective, especially when orchards were modest in size, and part of a balanced, mixed farming strategy set within biodiverse ecosystems. Not so long ago, sheep, pigs and fowls were often to be found under the big old apple trees – with the sheep keeping the grass down, the chickens pecking away at potential pests such as codlin moth and the pigs clearing up windfalls: the Gloucester Old Spot, once known as the "orchard pig", is said to taste better if it has been grazed in an orchard.[17]

There's now good evidence that communities of natural predators can sometimes function effectively as wild pest-control agents, especially where commercial pressures to satisfy supermarket demands for large quanitites of big and unblemished fruit can be avoided.

A head for heights: Ewan Greenhalf braves going up a ladder to harvest apples from a tall tree

In one remarkable study, simply putting up nestboxes for great tits significantly reduced caterpillar damage to the Dutch orchards investigated by two ecologists at the Netherlands Institute of Ecology at Heteren. The yield of apples was nearly doubled – from 4.7 kg to 7.8 kg per tree – in the orchards with a high density of breeding great tits.[18]

Another study, by government ecologists in Avignon in southeastern France, showed that great tits breeding in organic apple orchards were much more successful in rearing their young than were those nesting in conventional orchards or ones under an integrated pest management (IPM) regime, in which pest populations are monitored to fine-tune the timing of sprays.[19] The French researchers went on to show that the way orchards were managed had a powerful impact on many species of birds. Studying 30 species in all over three years in 15 orchards, the investigators found that bird abundance, species richness and diversity were all highest in organic orchards and lowest in conventional orchards.[20] As you'd expect, the conventional apple-growers' pest control strategy affected insect-eating birds more than seed-eating ones: the spray-wielding growers were damaging their allies, the insect-eating wild birds.

From the heartland of the American commercial apple industry, soil scientist John Reganold and his colleagues at Washington State University at Pullman have produced evidence that growing apples in organic systems can outperform conventional orchards "in environmental and economic sustainability".[21] Growing apples intensively in a bid to maximise yield is not necessarily the most sensible approach in the long view.

One reason old-style, non-intensive orchards can be surprisingly successful is that swathes of wildflower-rich grasslands underneath the trees help to ensure a ready supply of pollinating wild bees and hoverflies. The loss of these bee-friendly habitats in conventional orchards – where herbicides are routinely sprayed to wipe out competing vegetation under the dwarf trees – has contributed to a pollination crisis in many parts of the world. Honeybee hives have often been brought into orchards to try to compensate for the shortage of wild bumblebees and solitary bees, but even domesticated honeybees have recently suffered mysterious die-offs known as "colony collapse" in which exposure to pesticides has been implicated.[22] In the US and Japan, apple blossom has been laboriously pollinated by hand, or even mechanically in Japan, because insecticides and "clean" agricultural practices have virtually eliminated natural pollinators.[23]

Wild bees are now vital to the food security of our nation, say researchers at the University of Reading. According to research led by Professor Simon Potts of Reading's Department of Agriculture, Britain's honeybees are capable of pollinating at best only a third of the crops that rely on insect pollinators, including apples, strawberries, runner beans and oilseed rape. The economic value of pollination services to UK crops, estimated at £430 million per year, is now derived overwhelmingly from wild pollinators. Stuart Roberts, chairman of the Bees, Wasps and Ants Recording Society, adds: "Though many beekeepers still believe that honeybees are the most important pollinators, they can only pollinate a third of crops at most, and in reality they probably only contribute to 10-15 per cent of the work. Wild bees are the unsung heroes for our food security and so it is these species on which we need to focus our conservation efforts."[24] Bumblebees fly in colder weather than other bees and so are particularly important for pollinating early-flowering fruit trees and in cold years.

Low-carbon fruit-growing for the future

Climate change offers yet another good reason for planting apple trees, for there's now evidence to suggest that orchards can help to contribute to combat global warming. If orchards are planted in traditional style, and allowed to persist for many years, they effectively lock up carbon – thereby preventing its release into the atmosphere to fuel climate change – in both the trees themselves and in the untilled soil beneath them. A study of Herefordshire orchards by researchers at Bangor University recently concluded that low-intensity orchards with big, long-lived trees store significant amounts of carbon. Such orchards can tie up carbon as effectively as woodlands do.

Paradise for plantlife: orchards set within permanent grassland that is mown or grazed by livestock on a regular basis can develop a glorious diversity of wildflowers, such as (top to bottom) Self-Heal, Common Bird's Foot Trefoil and Meadow Cranesbill

And because apples grow well without the need for artificial fertilisers and the like, they turn out to have a low "carbon footprint" – a measure of the amount of greenhouse gases released into the atmosphere during the production of a product. According to data from Gareth Edwards-Jones of the University of Bangor, the typical carbon footprint of a kilogram of apples is a modest 0.5 kg of carbon dioxide, compared to 1.9 for broccoli, 2-5 for tomatoes and a whopping 12-23 kg of CO2 expended to produce a kilo of beef. Orchards are a "win-win – producing food and drink AND protecting against climate change".[25]

The benefits of planting fruit trees are also being promoted by the pioneers of "forest gardening" – building on traditional orcharding practices of intercropping with soft fruit or other crops, or grazing livestock or a flock of hens under the trees. Forest gardening, developed by permaculture specialists such as Martin Crawford at Dartington in Devon, involves growing a variety of useful trees, carefully chosen for food, fuel, fibre or timber, underplanted with a wide array of edible and medicinal plants as well as species useful for dyeing or fibre, or grown especially to attract bees and beneficial insects. Based on the natural ecology of a young temperate forest, the trees within the forest garden are carefully spaced to let light into the understorey layers. The results so far are impressive. Over less than 20 years, Martin has turned an ordinary two-acre pasture field into a remarkably self-sustaining and productive plot supporting more than 200 species in which fruit production plays a part.[26] Good examples in Sussex can be found in a private garden in Seaford and at Hertford Junior School, Brighton.

Although a full-blown forest garden requires a certain amount of research and planning, the basic idea of increasing our reliance on productive trees seems to be gaining ground once again – and not before time. A number of factors, including rising fuel costs, are likely to lead to increased in the costs of imports in the foreseeable future. In the 1950s, apple orchards were widespread in Sussex, offering food security on everyone's doorstep. What's more, ordinary folk knew a lot about apples, because they were commonplace, and taken-for-granted. That everyday familiarity with looking after particular sorts of apple trees in particular localities, knowing how to keep the fruit and how to cook with a great wealth of different varieties – all that vernacular understanding and local expertise is what real sustainability is all about, Common Ground has argued. The challenge now is to find ways to rebuild that practical wisdom once again, ready "for a time when local produce and varieties will be valued in new contexts".[27]

A good place to start might be our parks and squares and playgrounds, many of which could readily accommodate an apple tree or two, or more, as can many allotments and private gardens – and gardens cover some 9,000 hectares in East Sussex alone.[28] These green spaces constitute a largely untapped resource for fruit growing, in both urban and rural settings. And while many commercial growers under pressure to produce will not be able to envisage abandoning an array of pesticide sprays anytime soon, people growing apples on a smaller scale, perhaps just for their families and their friends, can adopt another way of orcharding that leaves plenty of room for wildlife.

At Wisborough Green in West Sussex, Sarah and Stuart Crabtree are looking after an old orchard alongside their house, keeping it in good heart and giving it a future. "We've got several 30-year-old trees, and one probably about 50 years old," Sarah says. The oldest tree grows bright red apples with a delicious raspberry flavour, which the Crabtrees have been able to store in their garage right through April. "Woodpeckers and tits make good use of the holes in the old tree, and we are trying to keep the orchard as traditional as we can – the land underneath hasn't been worked for 30 years and is full of wild plants." Electric netting keeps the chickens foraging in the orchard safe. With support from the West Weald Landscape project, the Crabtrees have also planted a new two-acre orchard of 78 trees of traditional local varieties. Some of the trees have been planted nine metres apart, on M25 rootstock, and so will grow into widely spaced, full-sized "standard" trees at least 11 feet (3.5 metres) high.

In the village of Barns Green, part of Itchingfield parish west of Horsham, Merve Goddard and family are looking after a small old orchard of about a third of an acre. By studying old maps, Merve has discovered that their orchard dates from at least the 1880s, when "Henleys", a substantial property with an impressive orchard, was established on old farmland. "This orchard was gradually eroded as the Henleys plot was subdivided and more houses were built. By the 1940s, four properties occupied the original plot. Today, all four of us have long gardens with the remnants of orchard at the bottom, but it is only really in ours that any substantial number of trees survive." After waging a sustained battle against a "mammoth bramble infestation" that was threatening the trees, the Goddards now have a stand of 10 fruit trees in good condition, "plus one head-height dead stump for the woodpeckers".

Prize-winners: these Crawley Beauty apples grew on a tree in a Sussex allotment

The biggest trees are very substantial, with trunks 50cm plus in diameter and heights of 8 to 10 metres. The orchard has two cooking apples, seven eaters and one pear tree. The cookers have been identified as Reverend W. Wilkes and Grenadier. "Both crop well and are fantastic in pies and crumbles or just cored and baked." The eaters include an Ida Red, and four as yet unnamed varieties – "one is a sort of russet and others more smooth skinned and pale green. All are good eaters, and crop well but ripen quite late," Merve says. "The last two years we have had a really good crop from our trees – far more than we can eat – so we have offered them free to the village and have had wine, chocolates and homemade cider in return – the barter system is alive and well in Barns Green."

Merve adds: "It has given me great pleasure getting our little orchard up and running again. I steer clear of pesticides and other chemicals and try and leave patches of rough grass and timber piles around and we get plenty of wildlife including foxes, deer, bats, hedgehogs, moles, lots of birds and plenty of insects. The orchard is providing for all of us – just as it should be!"

But you don't need to own land to have access to traditional orchards. Many allotments are peppered with fruit trees planted in earlier decades and some older orchards, such as the ones at Stanmer Park, are looked after by people who live in towns and cities, who have little or no garden space of their own. Their participation in the work days, courses, school and community visits in the orchards now means that there are many people who enjoy and appreciate the particular pleasures and benefits of spending time among fruit trees.

Not just for countryfolk: city-dwelling volunteers keep the Stanmer orchards in good heart, and vitally help with the harvest too

Furthermore, the new orchards being created at the heart of communities are redefining the notion of a traditional orchard. The scores of fruit trees planted on the Craven Vale housing estate in Brighton will enable hundreds of flat dwellers to look out upon and soon enjoy the fruits of trees bearing apples, plums, figs and damsons. In both urban and rural settings, plots of land next to or within residential areas of all kinds are beginning to be planted with fruit trees that will provide homes for a host of wildlife as well as food for the people who tend them and value their presence.

Endnotes for Chapter 11

1 www.ptes/org/orchards

2 Sue Clifford and Angela King, *Orchards – a guide to local conservation*, London: Common Ground, 1989; Sue Clifford and Angela King, *The Common Ground Book of Orchards*, London: Common Ground 2000, 16.

3 Sue Clifford, website, www.commonground.org.uk; www.england-in-particular.info

4 S. Baker, A.E. Burrough, L. Cordrey, K. Merry, and C. Wedge, 'Conserving the wildlife of traditional orchards', *British Wildlife*, 23 (2011), 8-16; http://www.ukbap.org.uk/library/BRIG/SHRW/orchards

5 Angela King and Sue Clifford, *Community Orchards Handbook*, Totnes: Green Books, 2011, 220.

6 F.W.M. Vera, *Grazing Ecology and Forest History*, Wallingford: CABI Publishing, 2000.

7 M.J. Smart and R.A. Winnall (eds), The biodiversity of three traditional orchards within the Wyre

Forest SSSI, Worcestershire: a survey by the Wyre Forest Study Group, *English Nature Research Reports*, No. 707 (2006).

8 M. Lush, H.J. Robertson, K.N.A. Alexander, V. Giavarini, E. Heewins, J. Mellings, C.R. Stevenson, M. Storey and P.F. Whitehead, Biodiversity studies of six traditional orchards in England, *Natural England Research Report* No. RINO25 (2009).

9 James Marsden, 'Apples don't have maggots – an overview of orchard wild life', in *Orchards and Wildlife*, report of symposium convened by Common Ground, 1999, 4-13.

10 Alexander, KNA and Bower, L, 'The Noble Chafer and traditional orchards – an old-growth species in the English cultural landscape', *British Wildlife* 23 (2011), 17-22.

11 Roger Key, 'Orchards and invertebrate biodiversity', in *Orchards and Wildlife*, report of symposium convened by Common Ground, 1999, 21-24.

12 Jonathan Briggs, 'Mistletoe-a review of its distribution, conservation and insect associates', *British Wildlife* 23(1): (2011), 23-31.

13 Natural England, Technical Information Notes on traditional orchards, TIN015, www.naturalengland.org.uk.

14 Michael Proctor, Peter Yeo and Andrew Lack, *The Natural History of Pollination*, The New Naturalist, London: HarperCollins, 1996, 362-3.

15 Petra Billings, 'Orchard heritage', *Wildlife* 165, September 2011, Sussex Wildlife Trust; Rachel Bicker and Petra Billings, 'The condition of traditional orchards in the West Weald Landscape area', Report from West Weald Landscape Partnership, Sussex Wildlife Trust 2011 website, www.swt.org.uk.

16 Kenneth Mellanby, *Pesticides and Pollution*, New Naturalist, London: Collins 1970, 2nd ed, 201-2.

17 Angela King and Sue Clifford, 'The Apple, the Orchard, the Cultural Landscape', in *Local Distinctiveness: place, particularity and identity*, London: Common Ground, 1993, 76.

18 Christel Mols and Marcel Visser, 'Great tits can reduce caterpillar damage in apple orchards' *Journal of Applied Ecology*, 39 (2002), 888-899.
C. Mols and M. Visser, 'Great tits (*Parus major*) reduce caterpillar damage in commercial apple orchards' *PloS ONE* 2(2) (2007), e202.

19 J.C. Bouvier, J.F. Toubon, T. Boivin, B. Sauphanor, 'Effects of apple orchard management strategies on the great tit (*Parus major*) in southeastern France' *Environ Toxicol Chem* 11 (2005), 2846-52.

20 J.C. Bouvier, B. Ricci, J. Agerberg, C. Lavigne, 'Apple orchard pest control strategies affect bird communities in southeastern France' *Environ Toxicol Chem* 1 (2011), 211-9.

21 John Reganold, Jerry Glover, Peston Andrews and Herbert Hillman, 'Sustainability of three apple production systems' *Nature*, 410 (2011), 926-929.

22 Alison Benjamin and Brian McCallum, *A World Without Bees*, London: Guardian Books 2008.

23 Proctor et al, op. cit., 363.

24 http://www.reading.ac.uk/news-and-events/releases/PR367212.aspx.

25 See website www.herefordorchards.co.uk/fls/climate.asp; Sarah Calcutt, "British apples cut carbon footprint", *South East Farmer*, May 2012, p 74.

26 www.agroforestry.co.uk; Martin Green, *Creating a Forest Garden*, Burlington: Chelsea Green, 2010.

27 King and Clifford, op. cit., 76.

28 Mike Toms, 'Sussex gardens: good for birds or birdwatchers?' *The Sussex Bird Report*, No. 61 (2008), Sussex Ornithological Society.

12. Finding Out More

12

Finding out more, ideas for action

Apples have been cultivated in Sussex for two thousand years, by Roman settlers, medieval peasants, monks, cottage dwellers, country landowners, vicars, allotment-holders, townspeople, farmers and professional fruit-growers. Fruits from other parts of the world with warmer climates enrich our diets, but many more of us could have ready access to local apples and other fruits: by growing our own in gardens, parks, school grounds and community spaces, harvesting fruit that currently goes to waste, supporting local growers, juice and cider makers, and learning how to cook and store fruit. We could be eating local fruit all year round and it could become a cornerstone in an enduring way of feeding ourselves well into the future. The focus of this book is on apples, but of course there are lots of other orchard fruits to enjoy: pears, plums, cherries, quinces, figs, mulberries, medlars, apricots, nut trees… and a whole range too of soft fruits – berries and currants – that thrive in Sussex soils. Here are a few practical ideas and sources of information and support for readers who would like to take action to ensure our fruit-growing future.

Grow your own fruit

You can plant apple trees in most conditions, if they are not too exposed to high winds and sea spray nor too shady. You can grow fruit trees even in quite small gardens, against fences or walls, so that they take up very little space but still produce lots of fruit. Buy your trees from a specialist nursery or garden centre; most nurseries will deliver the trees or send them by courier.

Choosing your trees

◊ There are many different types of eating and cooking apple that can be grown, each with a distinctive colour and flavour that will provide fruit from August until the following March.

◊ Most varieties will grow successfully in a wide variety of soil conditions and will provide fruit for many years.

The Sussex Apples tea towel was created by Erica Smith for Action In Rural Sussex using an illustration commissioned from Chris Watson

Spoilt for choice: hundreds of apple varieties are available from specialist nurseries. This watercolour shows Tinsley Quince, a sweet, firm apple with a faint quince-like flavour.

◊ Some varieties such as Discovery, Charles Ross, Ribston Pippin and Orleans Reinette show a natural resistance to fungal diseases so are less prone to scab and canker. Cox's Orange Pippin, however, is difficult to grow well and is susceptible to many diseases.

◊ The Sussex varieties (see chapter 2) also show a wide range of characteristics from the sweet Saltcote Pippin and aromatic Tinsley Quince to a sharper cooking apple like Alfriston or Forge. Some varieties such as Hawkridge ripen in early September, while Mareda is a late October apple. Crawley Beauty is the last apple variety to flower during early June, and because of this is always a useful variety in frost-prone areas.

◊ Before buying trees it is worth considering the size the tree will grow to and the pollination requirements of the variety.

◊ Apple trees are all propagated by grafting onto a rootstock. It is the rootstock which determines how big the trees will grow to and in some cases gives a resistance to pests such as woolly aphid and diseases such as crown rot.

◊ The dwarfing rootstocks such as M9 will produce a tree that can be grown in a smaller garden or as a cordon but will require a permanent post and feeding throughout its life.

◊ The intermediate size of rootstock, MM106 and M26, commonly sold in nurseries and garden centres, will grow well in most reasonably fertile soil and can also be grown as cordons and espaliers.

◊ More vigorous rootstocks such as M111 will produce a larger tree and are particularly useful because they will grow in more difficult conditions found on the less fertile soils or on the poorly drained clay of Sussex.

◊ These more vigorous rootstocks are also useful on the soils of the chalk downs which vary in depth from the deeper soils of the valleys and coombes where good conditions for apple growing can be found, to the thinner soils of the slopes and hilltops. Here, on these thin, dry soils, apples can be grown in the more sheltered places but will require mulching and feeding to grow well.

◊ If you want a tree that will one day provide you with enough shade under which to sit under and read a book, you can grow a "standard" tree on a very vigorous rootstock such as M25.

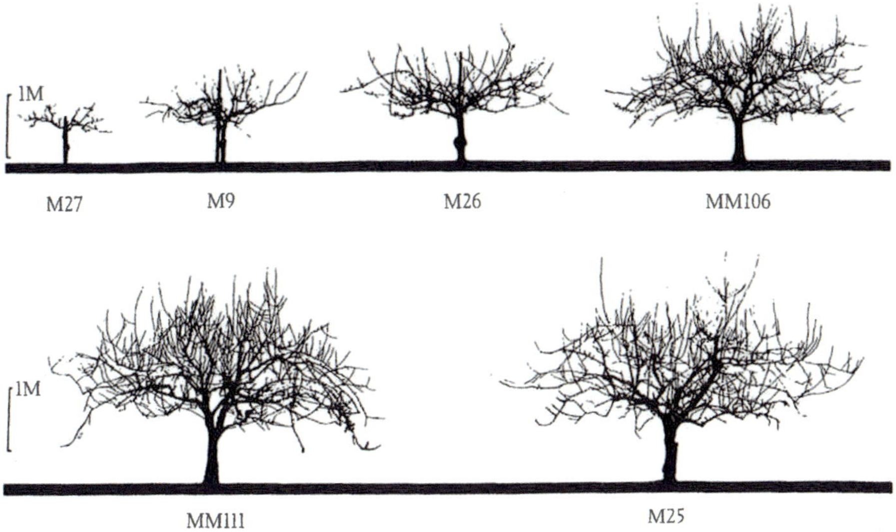

Dwarf to standard: rootstocks (named M or MM followed by number) determine how big an apple tree will grow

◊ Specialist nurseries offer apple trees grown on a wide range of rootstocks to suit the space and soils available.

◊ The trees are delivered from the end of November until early March when the trees are dormant and can be easily transplanted.

◊ Apple trees grown in containers and sold at garden centres can be planted throughout the year but they do need more watering in the early years to help them establish.

◊ Most apple trees will produce more fruit if they are planted with at least one other different variety that flowers at the same time. This proximity helps to ensure that the flowers are cross-pollinated and so set more fruit.

◊ One Sussex variety, *Crawley Beauty,* is self fertile and does not need a pollination partner, so can be planted as a single tree in a garden. In a suburban garden, however, there are often enough apple trees nearby for a tree to be adequately pollinated even if it is planted by itself.

◊ Consult a good gardening book (see below) for more information on pollination groups or look at the Keepers Nursery website, keepers-nursery.co.uk

For more information on apple varieties:
Joan Morgan and Alison Richards *The New Book of Apples* , London: Ebury Press, 2002, provides descriptions of more than 2000 apple varieties
Martin Crawford, *Directory of Apple Cultivars*, Agroforestry Research Trust, 46 Hunters Moon, Dartington, Totnes, Devon TQ9 6JT, agroforestry.co.uk
Michael Clark, *Apples, A Field Guide*, Stowmarket: Whittet Books, 2003

Specialist suppliers of apple trees, including Sussex varieties:
Bernwode Fruit Trees, Kingswood Lane, Ludgershall, HP18 9RB
bernewodefruittrees.co.uk
Blackmoor Nursery, Liss, Hampshire
blackmoor.co.uk
Brogdale Orchards Ltd, Brogdale Road, Faversham ME13 8XZ
brogdale.org
Brighton Permaculture Trust, Stanmer Park, Falmer
brightonpermaculture.org.uk
Herons Folly Garden, Fletching Street, Mayfield. TN20 6TE
heronsfollygarden.co.uk
Keepers Nursery, Gallants Court, East Farleigh, Maidstone, Kent ME15 OLE
keepers-nursery.co.uk
R.V. Roger, Malton Rd, Pickering, North Yorkshire, YO18 7JW
rvroger.co.uk
South Downs Nurseries and Garden Centre, Hassocks, West Sussex
south-downs.co.uk
Southern Fruit Trees, The Old Grain Dryer Corner, Blackmoor, Liss
blackmoorplants.co.uk

Good start: a perfectly planted young tree

Tree planting

This method of fruit-tree planting has been developed over a number of years and is now used in all the school and community orchards for which the Local Fruit Futures team is responsible.

◊ Fruit trees should be planted between the end of November and early March.

◊ It is best to avoid planting when conditions are very wet or the ground is frozen.

◊ Tree roots should not be allowed to dry out and must be kept covered and damp at all times. Soak the roots in a large container of water for an hour before planting.

◊ All trees should be staked at planting to keep them steady until the roots can grow and provide enough support.

◊ Use two stakes with the tree held by flexible ties in the middle (see diagram). This prevents the tree rubbing and allows it to move slightly in the wind.

◊ One stake with one tie is also perfectly acceptable for one-year old fruit trees.

◊ One-year-old apple trees should have 1.5m posts driven firmly into the ground, while larger trees will need 2.0m posts.

◊ Dig a hole wide and deep enough to ensure the roots are well spread out and covered.

◊ Remove the turf first and place it one pile, then dig out the top soil and place it in a separate pile. Finally, the sub soil is dug out and placed in a third pile – the sub soil is usually recognisable by its pale colour.

◊ The sides and base of the hole are then forked over. This aids drainage and allows the roots spread outwards.

◊ The two posts are then firmly driven into the ground 40cms apart. A post driver is useful for this.

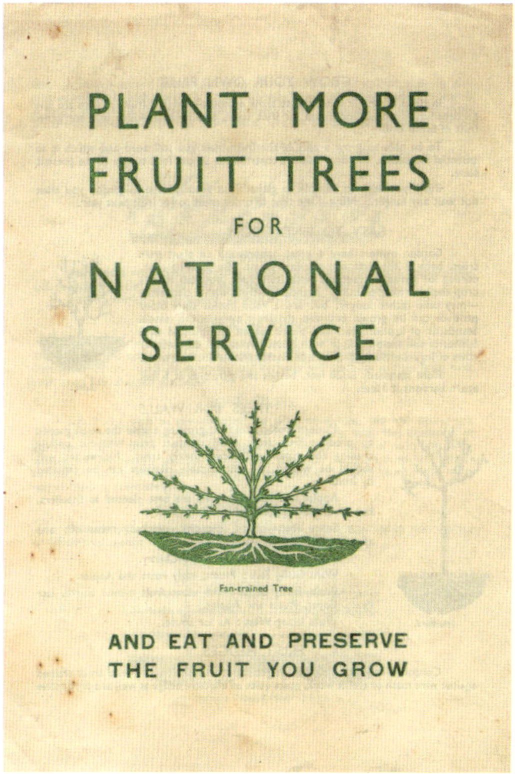

A public-sprited message from the war years still has resonance today

◊ Next, the turf is chopped up finely and placed at the bottom. The young tree is then positioned in the middle of the hole between the two posts.

◊ The graft union should be well above the soil level so that there is no possibility of the variety rooting into the soil which would neutralise the influence of the rootstock.

◊ Trees grown in containers will need to have the compost and roots loosened before planting, otherwise the roots remain tightly packed together.

◊ Add Rootgrow™, a mycorrhizal fungi, around the tree roots. This encourages the rapid establishment and vigorous growth of transplanted trees.

◊ Insert a perforated irrigation tube into the planting hole near the tree roots. Water goes to where it is needed near the roots.

◊ The topsoil is then replaced, treading down gently as the hole is filled so that soil makes good contact with the roots. A barrowload of compost can be mixed with the topsoil where soil conditions are poor. Fresh manure should not be used in the planting hole.

- ◊ Finally, the subsoil is replaced and a further check made to see that the graft union is well above the soil level.
- ◊ One tie is looped around the tree with a second positioned opposite to hold the tree between the two posts (see diagram).
- ◊ After planting mulch the base of the tree with chipped bark or straw to conserve soil moisture and suppress weeds.
- ◊ Alternatively a 1m square of black woven polypropylene can be pegged down around the tree. The base of the tree should be kept weed free for four years before allowing grass to re-grow around it.
- ◊ Wire or plastic mesh should be placed around the base of the tree to prevent damage from rabbits. The mesh also protects the tree from damage by strimmers.
- ◊ Finally, label the tree with the variety name and the date on which it was planted.

Safeguards for survival: planters at Fittleworth Primary School wrap plastic mesh around the tree to protect against rabbits, which gnaw fruit tree bark in winter, and accidental strimmer damage

Team work: students from Imberhorne Secondary School, East Grinstead look justifiably pleased with their morning's work – this is one of eight fruit trees they planted

Further information on fruit growing:
Keepers Nursery www.keepersnursery.co.uk.
Harry Baker, *RHS Growing Fruit*, Royal Horticultural Society, Ward Lock, 1999.
Bob Flowerdew's Complete Fruit Book, Kyle Cathie, 2000.

Buy local fruit

- ◊ At local shops – increasingly greengrocers are flagging up locally grown produce
- ◊ At farmers' markets and village markets
- ◊ From farm shops
- ◊ Pick Your Own
- ◊ In supermarkets – a few Sussex growers do supply supermarkets via wholesalers but the apples have to travel a circuitous route to get there, and are never (so far, in our experience) labelled as from Sussex. But you can at least look out for English-grown apples

Fresh from the tree: you can pick your own apples at Grange Farm, Funtington and many other places across Sussex

Here are some of the Sussex growers who sell some or all of their fruit through local outlets:

Durleighmarsh Fruit Farm, Rogate (PYO, farm shop)
durleighmarshfarmshop.co.uk
Grange Farm Shop, Funtington, (PYO, farm shop)
grangefarmshop.com
Greenway Fruit Farm, Herstmonceux (farmers' markets, local shops) 07843 576625
Laines Organic Farm, Cuckfield (local shops, farmers' markets) 01444 452480
Laurel Tree Fruit Farm, Crowborough (farm shop) 01892 661637
Maynards Fruit Farm, Ticehurst, (farmers' markets, PYO, farm shop, local shops) maynardsfruit.co.uk 07885 226452
Moat Farm, Salehurst (local shops, farmers' markets) 01580 880459
Oakwood Farm, Robertsbridge (local shops) 01580 830893
Oast Farm, Buxted (farm shop, PYO, local shops, farmers' markets) 01825 733446
oastfarm.com

Old Mill Fruit Farm, Bolney (farm shop, local shops, farmers' markets) 01444 881356
oldmillfarmshop.co.uk
Parsonage Farm, Udimore, (PYO and farmers' markets) 01797 222009
Paternosters Fruit Farm, Warninglid (PYO) 01444 461471
paternostersfarm.webs.com
Perryhill Orchard, Hartfield, (PYO, farm shop) 01892 77059
perryhillorchards.co.uk
Redcoat Orchard, Catsfield, (PYO, farm shop) 01424 772685
Ringden Fruit Farm, Flimwell (farm shop, farmers' markets, local shops) 01580 879385
ringdenfarm.co.uk
Roundstone Farm, Ferring (PYO, farm shop) 01903 783817
roundstonefarm.co.uk
Runcton Farm, Runcton (farm shop, local shops, farmers' markets), 01243 787847
Sharnfold Farm, Pevensey (PYO, farm shop) 01323 768490
sharnfoldfarm.co.uk
Stonehill Farm, Horam, (PYO, farm shop, local shops) 01825 872553
stonehillfarm.co.uk
Tablehurst Farm, Forest Row, (farm shop, local shops, farmers' markets) 01342 823173
tablehurstandplawhatch.co.uk
Tanyard Fruit Farm, Ninfield (PYO, farm shop, local shops, farmers' markets) 01424 892258
Tendrings Fruit Farm, Magham Down (farm shop, local shops) 01323 842116
Tullens Fruit Farm, Pulborough (farm shop, farmers' markets, local shops) 07967 110639
tullens.co.uk
Tulleys Farm, Turners Hill (farm shop, PYO) 01342 718472
tulleysfarm.com

Some orchards run by communities also sell some of their produce; Ringmer and Stanmer orchards for instance sell apples, juice and chutneys at events and sometimes through local shops.

Eat local apples for free

Many people with an apple tree or two in their gardens are happy to share their surplus and put out boxes of apples for passersby to help themselves.You might find abandoned or wilding apple trees happily growing along roadsides or footpaths, in parks and car parks, in hedgerows or at the edges of towns in waste land and in empty lots, with tasty apples free for the harvesting.

Picking apples from other people's trees without their permission is a time-honoured custom known as scrumping. The word has been borrowed and adapted by Brighton

Permaculture Trust, for its project to harvest and make use of apples that would otherwise go to waste. But they are picking, eating, cooking and juicing apples with the permission of the tree owners! The produce is then turned into juice, jams and chutneys and sold. There's a similar scheme based in Hastings and several others are developing as we go to print.

Or you might want to join (or set up!) a community orchard, where volunteers share in the harvest (see below).

Drink local apple juice and cider

Many growers in Sussex now make their own apple juice, and some is organic, such as Matthew Wilson's Oakwood Farm juice. Blends as well as single apple varieties are available, so you can try out different flavours. Community-run orchards are also turning surplus produce into juice. Heat juice gently with cinnamon or ginger for a warming winter drink. Cider makers too are on the increase – visit Middle Farm's English Cider Centre, near Firle in East Sussex to try their free samples (for more details, see chapter 9).

Hanna Shepard and Amelia Peñaloza Bur serving organic Sussex apple juice

Cook with apples

The Apple Source Book, by Sue Clifford and Angela King with Phillipa Davenport (Hodder and Stoughton, 2007), is packed with tempting culinary ideas – cooking with apples doesn't have to be dull or difficult. The versatility of apples makes them a fantastic food: ideal in chutneys, with tomato and onion, perhaps, fantastic in soups, with butternut squash, and of course a wonderful ingredient in all manner of puddings, pies and cakes. Simply stewing a tasty cooking apple with a couple of cloves and a few chopped dates makes a simple but delicious accompaniment with yoghurt, ice cream, porridge and breakfast muesli. Many apples are especially wonderful baked – cored with a few raisins and dates in the space left.

Overleaf are two easy and adaptable recipes.

Spiced Squash & Apple Soup

Robin van Creveld, Community Chef, has taught helpers how to make this soup at several of our orchard plantings, which generally take place on cold winter days – this soup has proved to be very popular and warming. Here is Robin's recipe.

For about 2 litres:
Ingredients
1kg squash of your choice e.g. Butternut or Crown Prince (peeled, de-seeded and chopped)
2 medium apples (peeled, cored and chopped)
2 onions (chopped)
2 cloves garlic (chopped)
1 litre reduced salt vegetable stock
1 tbsp olive oil
1 tbsp mild curry powder
salt and pepper

Hungry work: Burwash Cub Scouts chatting over lunch after a morning's orchard planting and apple cookery

Method
1. Heat the oil in a large pot and fry the onions until they start to brown.
2. Add the chopped apples and squash to the pot and cook for a further 5 minutes.
3. Add the garlic and curry powder and cook for a minute more.
4. Add the stock and bring to the boil.
5. Reduce heat to a fast simmer and cook for 15 minutes.
6. Blend the soup if you prefer a smooth consistency and season with salt and pepper.

Serve this soup with toasted sesame or pumpkin seeds and a swirl of cream or yoghurt

Ave & John's Apple Ginger Pickle (with or without chilli)

Avis Carter, who is an active member of Brighton Permaculture Trust and has made many very good chutneys from Brighton apples, says this is one of her most popular recipes.

3 lb/1.35 kg apples (any)
1 lb/450g onions
4 oz/115g crystallized ginger
1 1/2 lbs/675g dark muscovado sugar
1 tbs salt
1/2 pint/280mls malt vinegar, into which has been crushed, several cloves of garlic and chopped red chillis to taste. I suggest 4 cloves of garlic and one chilli to start. If more is required, add them to the pickle at any stage of cooking.

Peel, core and chop apples. Slice onions. Simmer all the ingredients very slowly until thick and very dark. Sterilize 4 or 5 1 lb/450g jars and fill whilst the jars are still hot. Seal the lids. This pickle keeps well if the jars are properly sterilized.

NB you can sterilize jam jars by putting them (lids off) in an oven on a low setting for 15 mins or so.

Store apples over the winter

Storing fresh apples is a practice that is rare among young people, but still fairly common among people of earlier generations who have access in the autumn to an abundance of apples. Apples must be in good condition with no bruising or signs of damage from insects and preferably with the stalk intact. The fruit should be kept in cool, frost free, dark conditions away from mice – people use sheds, lofts, garages, cellars. Lay them out, not touching, and wrapped in newspaper or left unwrapped in slatted apple racks which will allow air to circulate around them. You can also use recycled mesh plastic trays that are used to display food in shops. It is best to check over the apples regularly and remove any rotten fruit. Apples can also be stored in plastic bags in the the fridge though the bags must be perforated so that some air can circulate around the fruit.

Winter store: many varieties of apples will last for months in a cool, dry place and grow sweeter as they age

Varieties of apple which ripen early such as Discovery or Hawkridge should be eaten within a few weeks, but some apples will keep for months. Of the Sussex varieties, Alfriston, a cooker, keeps very well and that's why it was taken on board ship for long voyages. First and Last also stores well (the clue presumably is in the name). Egremont Russet acquires more flavour and sweetness and a softer texture if stored.

Contrast and compare: named apples act as a handy reference to aid apple identification by Joan Morgan at Apple Day at Stanmer Park

Identify your fruit

Do you have an apple tree but don't know what it is? Apple varieties number in the thousands so apple identification is a highly specialised activity.

Brogdale Farm, home of the British National Fruit Collection in Kent, will identify apples (£10 per sample). Brogdale Farm, Brodgale Road, Faversham, Kent ME13 8XZ 01795 536250, brogdalecollections.co.uk

The Royal Horticulture Society hold fruit identification days during the summer and autumn at the four main RHS gardens at Harlow Car, Hyde Hall, Rosemoor and Wisley, rhs.org.uk

Apple Day celebrations such as Apple Day at Stanmer Park and the Apple Affair at West Dean Gardens often have an expert on hand to help identify visitors' apples.

Create (or join) a community orchard

Brighton Permaculture Trust has a wealth of experience in supporting people through the process of designing, planting and maintaining community and school orchards. It also runs one-day courses at Stanmer Park on setting up orchards, looking after young fruit trees and a whole range of other orchard skills including grafting, pruning and the use of scythes.
www.brightonpermaculture.org.uk

Plumpton College run courses on Planning and Planting an Orchard through the Centre for Sustainable Food, Farming and Forestry. 01580 879547,
foodfarmingforestry.co.uk

In West Sussex, Chichester District Council offers help and advice and funding: contact the Environment Officer, East Pallant, Chichester PO19 1TY 01243 785166. A *Guide to Setting Up Community Orchards* is also available.
chichester.gov.uk.

For a good all-round guide to setting up a community orchard look out for *The Community Orchards Handbook*, produced by Common Ground, reprinted by Green Books, 2011.
commonground.org.uk

You might want to visit or support one of the many Community Orchards already thriving in Sussex. The Home Farm Orchard at Stanmer Park, near Brighton, offers a local example of an old orchard restored and extended by a range of voluntary organisations. It is now fully accessible and is open to school and community group visits and runs orchard tours at various times of the year, not least at the annual Apple Day celebration. It is used by Brighton Permaculture Trust for its many courses on orchard skills. Ringmer Community Orchard in East Sussex has open days and workshops throughout the year, organised with the help of Common Cause, based in Lewes.
commoncause.org.uk.

In a more urban setting, Worthing now has its own Community Orchard too, established by its Transition Town team as do Hassocks and Chichester. Your local horticultural society may also be able to help.

Brogdale Horticultural Trust, Brogdale Road, Faversham, Kent ME13 8XZ offers short courses and open days.
brogdalecollections.co.uk

Many other organisations offer training courses and workshops including the Orchard Skills Centre in Gloucestershire.
dayscottage.co.uk

More information and support for local orchard groups available from:

The Orchard Network
orchardnetwork.org.uk

Orange Pippin
orangepippin.com

Apple art: apple picking is the theme of an embroidery by Caroline Zoob

Orchard skills: courses in scything, grafting and all aspects of fruit tree care are run by Brighton Permaculture Trust

Save old orchards

If you're lucky enough to have inherited old apple trees, they'll need careful handling. For guidance on how to look after old orchards gently, consult the very useful Natural England's technical information guides, available by post or on the web, at *naturalengland.org.uk*

The Orchard Network, launched in 2010 on behalf of the Traditional Orchard Biodiversity Action Plan, covers a range of topics from orchard care and management to grants and planning issues, at *orchardnetwork.org.uk*

Brighton Permaculture Trust runs training days on the pruning of mature fruit trees each year and can also offer site visits and advice on the restoration of old or neglected fruit trees. Brinsbury College also has a pruning day course.

You could join (or set up) a project to map local orchards and fruit trees, searching for survivors in city and country, such as the planned Lewes Abundance Project *friends-of-Lewes.org.uk*

Common Ground has been working to save old orchards since the 1980s. For more ideas on community action, see
commonground.org.uk and *england-in-particular.info*

There's also helpful information from the People's Trust for Endangered Species, phone 020 7498 4533 or
ptes.org

In West Sussex, the West Weald Landscape Project run by Sussex Wildlife Trust provides free advice to landowners with old orchards, see
swt.org.uk

You can try out ideas for growing fruit sustainably and encouraging wildlife in your orchard, with tips from the websites above, and from:

Garden Organic, Coventry
gardenorganic.org.uk
Permaculture Association
permaculture.org.uk

The Agroforestry Research Trust
agroforestry.co.uk
and see Martin Crawford's *Creating a Forest Garden: working with nature to grow edible crops*, 2010, Chelsea Green Publishing.

Colin Tudge
campaignforrealfarming.org

Older fruit trees can be restored to good health and made more fruitful by careful pruning

Take part in or organise Apple Celebrations

Apple Day

Celebrations of the apple harvest in Sussex are held every autumn at Stanmer Park, near Brighton, West Dean Gardens, near Chichester, Middle Farm, Firle and at many other places with orchards. There are usually opportunities to see and taste unusual apple varieties, drink freshly pressed apple juice, get advice on fruit growing, watch cookery demonstrations, and enjoy apple-based foods as well as apple games and entertainments. Many schools with orchards now celebrate the apple harvest by cooking and juicing local apples, putting children in touch in a very enjoyable way with the seasonality of local food sources.

Ideas in abundance for apple-themed celebrations and games are packed into Common Ground's *Apple Games and Customs*, 1994, reprinted 2005, available from Common Ground, Gold Hill House, 21 High Street, Shaftesbury, Dorset SP7 8JE.

Wassailing

Early January is traditionally the time for a ritual known as wassailing, which often involves a night-time visit to an orchard, making a "hullabaloo", anointing an apple tree with cider and singing wassailing songs - revelries designed, it is said, to ensure a good apple crop in the following season. The wassail at Old Mill Farm in Bolney is led by Morris Dancers and revellers carrying burning torches, and equipped with saucepan lids and trumpets, definitely make a loud enough noise to drive out any demons present.

Acknowledgements

The writing of this book has been a collaboration. Brian Short, Emeritus Professor, University of Sussex is a historical geographer and he is responsible for much of the detailed research on the history of our orchards. Peter May, an orchard expert, gathered the information on Sussex apples, interviewed growers and provided horticultural advice. Gail Vines, science writer and editor, made major contributions to the sections on orchards and wildlife and edited the text and Anne-Marie Bur, Co-ordinator: Food and Health, Action In Rural Sussex, devised the book, acted as picture researcher, drafted and amended parts of the text and kept the show on the road...

An enormous number of people have contributed to Apples and Orchards in Sussex with their stories, images, advice, information, technical skills and support. Many of them are credited in the text. In addition we would like to thank Teresa Gittins, Emma Drew and Steve Roberts, Action in rural Sussex, Jonathan Douch, cartographer, Steve Oram and Anita Burrough of the People's Trust for Endangered Species, John Bleach, Judy Brent and Chris Milburn of Sussex Archaeological Society, Frances Rankine of the V&A Museum, Christopher Whittick, East Sussex Record Office, Mary Pennell, National Fruit Collection at Brogdale, Jill Arnott of Hampshire County Archives, Tom Schaefer of the Helen Allingham Society, Lionel French, Clerk to the Worshipful Company of Fruiterers, Jim and Sarah Buckland of West Dean Gardens, Rev. Canon Donald Johnson, Alan Chapman, Paula and John Atkins formerly of Tullens Fruit

Farm, Topsy Jewell and Katharine Finnigan of Common Cause and Amanda Solomon of Sussex Wildlife Trust. We must add our special thanks to photographer Anne-Katrin Purkiss for her patience, diligence, good humour and stamina in the face of requests for photos to be taken, tracked down or edited usually at short notice and at all times of the day and night!

We would also like to thank Bryn Thomas of Brighton Permaculture Trust, fruit-growing expert, co-manager of Local Fruit Futures, an inspiration over many years and supplier of crucial advice and information at vital moments!

We are grateful too to all those who funded us to carry out the whole Local Fruit Futures project of which this book is a part; principally the Big Lottery Local Food Fund, who as well as financing our work have been constructive and supportive in all their dealings with us. We also extend thanks to Brighton and Hove City Council, East Sussex County Council, West Sussex County Council, and the South Downs Joint Committee (pre-cursor to the South Downs National Park) for their contributions.

Finally we would like to gratefully acknowledge the role played by all the people, past and present – the cidermakers, juicers, cooks, shopkeepers and costermongers, the gardeners, growers, and the pickers - who have made this land fruitful.

Pictures

All reasonable attempts have been made to trace and acknowledge copyright holders of images shown. We would welcome contact from any we have not managed to identify

Cover illustration: Michael Munday

Chapter One p.10 Anne-Katrin Purkiss; p.11 Kate Greenhalf; p.12 Nicky Ashworth; p.13 not traceable; p.14 AKP; p.15 Amberley Museum, courtesy of Peter Waller; p.16 Hedvig Murray; p.17 KG; p.18 AKP; John Webley; p.19 AKP; p.20 Sussex Archaeological Society; p.21 David Morgan; Brighton and Hove City Council

Chapter Two p.24 Nicky Ashworth; p.25 Michael Munday; p.26 MM; p.30 MM; p.32 AKP; p.33 MM; p.34 The National Fruit Collection at Brogdale, AKP, The National Fruit Collection at Brogdale; p.35 Peter May, The National Fruit Collection at Brogdale, The National Fruit Collection at Brogdale; p.36 PM, AKP, AKP; p.37 The National Fruit Collection at Brogdale, PM, AKP; p.38 AKP, The National Fruit Collection at Brogdale; p.39 AKP, The National Fruit Collection at Brogdale, PM; p.40 The National Fruit Collection at Brogdale, AKP, AKP; p.41 AKP, AKP; p.42 AKP, AKP, AKP; p.43 AKP, The National Fruit Collection at Brogdale, PM

Chapter Three p46 ESRO FRE 6909; p.47 MM; p.48 Barrie Juniper; p.50 © Rijksmuseum, Amsterdam; p.51 Sussex Archaeological Society; Brian Short; p.52 AKP; p.55 Worshipful Company of Fruiterers; p.56 Kent History and Library Centre

Chapter Four p.60 Royal Horticultural Society, Lindley Library; p.61 MM; p.62 Andy Gammon; p.64 © British Library; p.65 © British Library; p.66 RHS Lindley Library; p.67 RHS Lindley Library; RHS Lindley Library; p.68 RHS Lindley Library; p.70 Nicky Ashworth; p.72 © British Library

Chapter Five p.74 British Library; p.77 East Sussex Record Office GLY 914; p.78 © British Library; p.79 © British Library; p.80 MM; p.81 © National Portrait Gallery, London

Chapter Six p.86 ESRO Ref TD/E 12/1; p.87 MM; p.88 AKP; AKP; p.89 © RHS Lindley Library; p.90 © Estate of Charles Jones 2012. All Rights Reserved, DACS, and © Victoria and Albert Museum; p.91 © Estate of Charles Jones 2012. All Rights Reserved, DACS, and © Victoria and Albert Museum; p.93 Jonathan Douch cartography; p.97 Buish 1912 / Helen Allingham Society; p.98 Henry Herbert La Thangue, Cider apples, circa 1899, oil on canvas, 111.8 x 96.5cm Purchased 1913, Collection: Art Gallery of New South Wales; p.99 A Sussex Orchard by Henry Herbert La Thangue, Image thanks to the Art Renewal Center® www.artrenewal.org

Chapter Seven p.102 ESRO OS 32-13-1939 and OS 45-1-1939; p.104 Courtesy of Rory Clarke; p.105 Courtesy of Rory Clarke; p.106 Jonathan Douch; p.107 Ian Tyas / Hulton Archive / Getty Images; p.108 Museum of English Rural Life; p.109 MERL; p.110 Jonathan Douch; p.113 © Manchester Art Gallery; p.114 ©Hampshire County Council Arts & Museums Service; p.115 Hardiman Scott; p.116 MERL; p.117 David Atkins, p.118 Courtesy of Paula Atkins; p.121 Courtesy of Fiona Durling; p.122 Courtesy of Fiona Durling; p.123 Jonathan Douch; p.125 MM

Chapter Eight p.129 MM; p.131 AKP; p.132 Courtesy of Rory Clarke; p.134 AKP; p.136 PM; p.139 AKP; p.140 Courtesy of Sally Cave (both images); p.141 Courtesy of Gladys Marshall; p.142 AKP; p.143 AKP; p.144 Courtesy of Tom Mersul; p.145 Courtesy of Tom Mersul; p.146 Courtesy of Fiona Durling; p.147 AKP; p.149 AKP; p.150 Courtesy of Graham Love; p.151 AKP; p.155 AKP; p.156 AKP; p.158 AKP; p.160 AKP

Chapter Nine p.162 AKP; p.163 MM; p.167 AKP; p.168 AKP; PM; p.172 PM; p.173 Courtesy of Perryhill Orchard; p.174 AKP; p.175 Anne-Marie Bur; p.176 AKP

Chapter Ten p.178 AKP(3), KG; p.179 MM; p.180 Katharine Finnigan; p.181 AMB; AKP; p.182 Courtesy of Grays Infant School; AKP; p.183 Courtesy of Sarah O'Dair; © BHCC; p.184 PM; Nicky Ashworth; p.185 PM; p.187 PM; p.188 Courtesy of Castledown Primary School (both); p.189 AKP; p.190 AKP; KG; p.191 AKP

Chapter Eleven p.192 AKP; p.193 MM; p.194 People's Trust for Endangered Species(c) Crown copyright 2012. Ordnance Survey Licence number 100022021,Data: Natural England (c) Crown copyright. All rights reserved 2012 ; p.195 PM; © Colin Smith; p.196 AKP; p.197 AKP; AKP; p.198 © Lisa Geoghegan, Sussex Wildlife Trust; Paul Brock; p.199 © John Webley; © Derek Middleton, Sussex Wildlife Trust; © John Webley; © John Webley; © John Webley; p.200 AKP; AKP; © Derek Middleton, Sussex Wildlife Trust; © Alan Price, Sussex Wildlife Trust; © Derek Middleton Sussex Wildlife Trust; p.201 Jo Simmons; AKP; p.202 KG; p.203 KG; p.204 PM; PM; AKP; p.206 Gail Vines; p.207 Brighton Permaculture Trust

Chapter Twelve p.210 © Chris Watson; p.212 Nicky Ashworth; p.213 Courtesy of East Malling Research; p.214 David Bradford; p.215 The British Growers' Publicity Council; p.216 Fittleworth Primary School; Imberhorne Secondary School; p.212 AKP; p.219 AKP; p.220 AKP; p.221 AKP; p.222 AKP; p.223 Caroline Zoob; p.224 Brighton Permaculture Trust; PM; p.225 Brighton Permaculture Trust

Index

A

B

C

D

E

F

G

H

I

J

K

L

M

N

O

P